Selection Indices and Prediction of Genetic Merit in Animal Breeding

Selection Indices and Prediction of Genetic Merit in Animal Breeding

N.D. Cameron
Roslin Institute
Edinburgh
UK

CAB INTERNATIONAL

CAB INTERNATIONAL
Wallingford
Oxon OX10 8DE
UK

Tel: +44 (0)1491 832111
Fax: +44 (0)1491 833508
E-mail: cabi@cabi.org

CAB INTERNATIONAL
198 Madison Avenue
New York, NY 10016-4341
USA

Tel: +1 212 726 6490
Fax: +1 212 686 7993
E-mail: cabi-nao@cabi.org

©CAB INTERNATIONAL 1997. All rights reserved. No part of this publication may be reproduced in any form or by any means, electronically, mechanically, by photocopying, recording or otherwise, without the prior permission of the copyright owners.

A catalogue record for this book is available from the British Library, London, UK
A catalogue record for this book is available from the Library of Congress, Washington DC, USA

ISBN 0 85199 169 6

Printed and bound in the UK by Biddles Ltd, Guildford and King's Lynn from copy supplied by the author

Contents

Preface and Acknowledgements		viii
1	**Introduction to Variance**	1
	Introduction	1
	Properties of variance	3
	Analysis of variance (ANOVA)	6
	Repeatability	10
2	**ANOVA in a Quantitative Genetics Framework**	12
	Variance components	13
	Precision of estimated parameters	15
	Genetic interpretation of variance components	17
	Heritability	19
	Maternal effect	19
3	**Regression and Correlation**	21
	Covariance	22
	Correlation between traits	24
	Genetic interpretation of regression and correlation coefficients	27
	Genetic correlation between traits	28
	Repeated measurements	30
	Half-sib correlation	31
4	**Identification of Animals of High Genetic Merit**	33
	Information on the animal	34
	Selection differential	36
	Response to selection	38
	Variance of predicted breeding value	39
	Accuracy of predicted breeding value	40
	Prediction error variance	41

5	**Information from Relatives**	44
	Information from sibs	44
	Information from progeny	50
	Information from parents	52
	Responses with measurements on the animal, sibs and progeny	54
6	**Selection Index Methodology**	58
	Selection objective and selection criterion	58
	Selection criterion coefficients	60
	Responses to selection	62
	Contribution of traits in the selection objective	65
	Contribution of traits in the selection criterion	65
7	**Examples of Selection Objectives and Criteria**	68
	Measurements on an individual and the mean of its sibs	69
	Measurement of two traits on the individual	72
	Traits in the selection criterion but not in the selection objective	73
	Restricted selection objective	75
	Desired gains selection objective	79
	Example from a pig breeding programme	81
	Selection on several traits with information from relatives	86
	Economic values	90
8	**Factors Affecting the Rate of Genetic Improvement**	93
	Errors in genetic and phenotypic parameters	93
	Modification of parameter estimates	95
	Generation interval	98
	Reduction in variance (Bulmer effect)	99
	Inbreeding	101
9	**Performance Testing, Progeny Testing and MOET**	106
	Optimising performance testing and progeny testing	106
	Progeny testing and MOET (multiple ovulation and embryo transfer)	109
10	**Simultaneous Prediction of Breeding Values for Several Animals**	114
	Sire evaluation	115
	Animal evaluation	118
11	**Prediction of Breeding Values and Environmental Effects**	121
	Best linear unbiased prediction (BLUP)	122
	Sire model	123
	Repeatability model	129
	Individual animal model	130
	Prediction error variance	133

12	**Multivariate Breeding Value Prediction**	135
	Equal design matrices	135
	Predicted genetic merit with unmeasured traits	136
	Common environmental effect	137
	Maternal genetic effect	138
	Unequal design matrices	140
	Genotype with environment interaction	140
13	**Breeding Values with a Gene of Known Large Effect**	143
	Parameterising a gene of known large effect	143
	Genotypic values and breeding values	144
	Selection with a gene of known large effect	146
	Responses to selection	148
	Genetic markers	150
	Breeding values with marker assisted selection	151
14	**Breeding Values for Binary Traits**	155
	Threshold model and liability	155
	Response to selection	157
	Heritability estimation	159
	Linear and non-linear models	163

References 165

Appendix
- Matrix algebra — 171
- Normal distribution tables — 175

Questions 176

Answers 183

Index 201

Preface and Acknowledgements

The objective of this book is to describe methods of identifying animals of high genetic merit, for characters of interest, subsequently to be used as parents of the next generation in a genetic improvement programme. Information on an animal's own measurements, or the measurements on its sibs, progeny, parents or other relatives, can be used to predict an animal's genetic merit. Rather than being limited to measurements on one trait, information from several traits can be combined for the prediction of overall genetic merit. Given a particular method of predicting animals' genetic merit, the rate of genetic improvement can be determined for evaluation of alternative breeding programmes. Therefore, knowledge of the methods outlined in the book is fundamental for efficient management of genetic improvement programmes.

The book consists of five sections. In the first section consisting of Chapters 1–3, the statistical procedures and quantitative genetics theory necessary for predicting genetic merit are described. Readers with experience in estimation of genetic and phenotypic parameters can omit these chapters. However, regression of predicted genotype on phenotype is used to derive the selection index coefficients, so readers should be familiar with the ideas in Chapter 3. Prediction of genetic merit based on measurements on the individual and its relatives is discussed in Chapters 4–6, which form the second section of the text. Practical examples of breeding objectives and methods of predicting genetic merit are discussed in the third section, consisting of Chapters 7–9. The fourth section comprises Chapters 10–12, and outlines methods of predicting genetic merit, while simultaneously accounting for environmental effects, which affect animal performance. Chapter 10 uses selection index procedures to introduce the BLUP method, described in Chapters 11 and 12, which is used internationally within animal genetic evaluation organisations. In the fifth section, prediction of genetic merit when there exists a gene of large effect on the trait of interest is discussed in Chapter 13, followed in Chapter 14 by a similar discussion for a trait of a binary nature. A comprehensive list of references complements the text. In general, papers referred to in the text were chosen to reflect the present state of research in each subject, but reference to the original papers can be obtained from cited papers. There are examples with worked answers, throughout the text, to help the reader's understanding and a series of questions with detailed suggested answers completes the text.

The text originated from material used to teach Quantitative Genetics courses at Cornell University and the University of Edinburgh. At John Pollak's invitation, I taught Quantitative Genetics to final year Animal Science students at Cornell, which was the catalyst for writing this book. The experience of teaching the courses has helped my understanding of Quantitative Genetics, as has the interaction with the students. Charlie Smith and Robin Thompson both taught me a great deal about Animal Breeding and Quantitative Genetics, which I hope is reflected in the book. Parts of the text were written at Cornell and Edinburgh, but it was in Guejar-Sierra, in the Sierra Nevada mountains, that the first draft was completed. Bill Hill read all of the manuscript and made many helpful suggestions, for which I am very grateful. Finally, I would like to thank Vicky Cameron for the continual encouragement, interest and patience that she has shown during the writing of this book.

Chapter one

Introduction to Variance

Introduction

Genetic improvement is the process of selecting animals of higher genetic merit than average, to be parents of the next generation, such that the average genetic merit of their progeny will be higher than the average of the parental generation. Already, terminology has been introduced, which is not used in everyday language. What does the phrase "selecting animals of higher genetic merit" mean and how is "average genetic merit of their progeny" measured? It is necessary to be familiar with the concepts used in the practice of genetic improvement, before different methods of genetic improvement can be examined.

The first three chapters provide the statistical and quantitative genetics background required for prediction of genetic merit and examination of alternative selection strategies. For further information, the reader is referred to Snedecor and Cochran (1989) and Falconer and MacKay (1996).

Population and sample

A population is defined as a group of individuals with a particular specification. Examples of populations are all Holstein–Friesian cows in Wisconsin or all Blackface ewes in Scotland. Obviously, every animal in the population cannot always be measured. Measurements can be made on a sample of the population to gain information about the population. The choice of sample is important, as animals in the sample should be representative of animals in the population. Examples of samples, given the above populations, would be first lactation Holstein–Friesian cows in Wisconsin or the Blackface ewes on three farms. Drawing conclusions about the population from the sample is more likely to be reliable in the second example than in the first, provided that the farms were chosen as being representative of different aspects of Scottish sheep farming. Therefore, it is important that the population is specifically defined, such that an appropriate sample of animals can be measured.

Mean and variance of a sample

The mean or average value of a sample provides information on where the observations or measurements are. For example, the mean growth rate of a particular group of animals was 800 g/day. The mean of the measurements made on N animals is

$$\overline{X} = \frac{1}{N}\sum_{i=1}^{N} X_i$$

where X_i is the measurement on the ith animal. To complement the mean, the variance provides information on the variation between measurements, to indicate the "spread" of values. The variance of N observations in a sample from the population is

$$s^2 = \frac{1}{N-1}\sum_{i=1}^{N}(X_i - \overline{X})^2$$

The variance is a measure of the average squared difference between each observation and the mean. If the squared term was omitted, then the average difference between the observations and the mean would just be zero.

The variance is measured in squared units, such as 25 mm^2, as squared terms are used in calculation of the variance. The standard deviation is the square root of the variance and can also be used as a measure of the variation between the observations. For example, in two groups of animals, the standard deviations for growth rate were 80 g/day and 140 g/day, indicating a relatively greater range of growth rate for animals in second group. The standard deviation can be preferable to the variance for expressing the range of observations, as the dimensions of the mean and the standard deviation are the same.

The variance can be calculated from the sum of the squared observations and the sum of the observations, squared, rather than calculating the square of each deviation from the mean:

$$\frac{1}{N-1}\sum_{i=1}^{N}(X_i - \overline{X})^2 = \frac{1}{N-1}\left[\sum_{i=1}^{N} X_i^2 - N\overline{X}^2\right] = \frac{1}{N-1}\left[\sum_{i=1}^{N} X_i^2 - \frac{1}{N}\left(\sum_{i=1}^{N} X_i\right)^2\right]$$

For example, weights of a sample of five animals from a herd were 13, 15, 11, 14 and 12 kg. The variance is calculated from the sum of the squared observations

$$\sum_{i=1}^{n} X_i^2 = 13^2 + 15^2 + 11^2 + 14^2 + 12^2 = 855$$

and the sum of the observations squared

$$\left(\sum_{i=1}^{n} X_i\right)^2 = (13 + 15 + 11 + 14 + 12)^2 = 4225$$

and the variance is $\frac{1}{4}\left(855 - \frac{1}{5}4225\right) = 2.5$.

When calculating the mean and the variance for a sample of size N, the divisor for the mean is N, while the divisor for the variance is N–1. The difference in divisors comes about because calculation of the variance requires information on the mean, such that there are only N–1 independent observations, once the mean is known. For example, the mean measurement of five animals was 3.6, such that if the measurements on the first four animals, X_1, X_2, X_3 and X_4, are known, then the fifth observation must equal $5(3.6) - (X_1 + X_2 + X_3 + X_4)$.

Different notation is used for the mean, variance and standard deviation of the population and of the sample, to indicate the source of the parameters, as shown in Table 1.1.

Table 1.1. Notation for population and sample parameters

	Population	Sample
Mean	μ	$\overline{X}$
Variance	σ^2	s^2
Standard deviation	σ	s

Properties of Variance

The variance has properties that are used throughout the text, and it is important that these properties are understood at the beginning. The variance of trait X will be denoted by var(X) and α is a constant.

Property 1 $\text{var}(X + \alpha) = \text{var}(X)$
Adding the constant α to each observation does not change the variance, as the variation between observations remains the same, although the mean changes by the amount α. If the variance does not change, then neither does the standard deviation.

Property 2 $\text{var}(\alpha X) = \alpha^2 \text{var}(X)$
Multiplying each observation by the constant α increases the variance by α^2. The variance was calculated from the squared deviations from the mean, so for each observation $(\alpha X - \alpha \overline{X})^2 = \alpha^2 (X - \overline{X})^2$, such that the variance increases by α^2. If the variance increases by α^2, then the standard deviation increases by α. The mean will also increase by α.

Property 3 If X and Y are two independent or unrelated traits, then:
(a) $\text{var}(X + Y) = \text{var}(X) + \text{var}(Y)$;
(b) $\text{var}(X - Y) = \text{var}(X) + \text{var}(-Y) = \text{var}(X) + (-1)^2 \text{var}(Y) = \text{var}(X) + \text{var}(Y)$;
(c) $\text{var}\left(\sum_{i=1}^{N} X_i\right) = N \text{var}(X)$, from 3(a) assuming equal variance for each X_i and all the X_i are independent.

Property 4 $\text{var}\left(\dfrac{1}{N}\sum_{i=1}^{N} X_i\right) = \dfrac{1}{N^2} \text{var}\left(\sum_{i=1}^{N} X_i\right) = \dfrac{1}{N} \text{var}(X)$

The fourth property of the variance is obtained by combining property 3(a) with property (2), with the constant α replaced by $\frac{1}{N}$. The fourth property translates into "the variance of the mean of N observations is $\frac{1}{N}$th of the variance of the observations". The variance of the mean, estimated from a sample of N observations, is a measure of the precision of the mean, as a mean with a smaller variance indicates a more precise estimate than a mean with a larger variance. The precision of a sample mean is a measure of the reliability of conclusions which can be made about the population mean, assuming that the sample is representative of the population.

The square root of the variance of measurements in a sample of size N is the standard deviation. However, the square root of the variance of the sample mean is the standard error of the mean. Therefore

$$\sqrt{\text{var}(X)} = \text{standard deviation}$$

$$\sqrt{\text{var}(\overline{X})} = \sqrt{\frac{\text{var}(X)}{N}} = \text{standard error}$$

The standard error of the mean decreases as the square root of the number of observations in the sample increases. For example, if the population standard deviation is 3 mm, then standard errors of sample means with 100 or 400 measurements will be 0.3 or 0.15 mm, respectively, such that quadrupling the number of measurements halves the standard error of the mean.

Normal distribution

Many traits are assumed to be normally distributed (Fig. 1.1), such that properties of the normal distribution are used for a variety of purposes. The normal or Gaussian distribution can be defined by the height of the normal distribution curve

$$\frac{1}{\sqrt{2\pi}} \exp\left(-\frac{1}{2}x^2\right)$$

for a point x standard deviation units from the mean. The normal curve is symmetrical about the mean (Fig. 1.1), as the formula for the height of the normal distribution curve is a function of x^2. Therefore, a specific proportion of the population is expected to deviate from the mean by a specific number of standard deviations.

For example, 0.34 of the population are expected to have observations between the mean and one standard deviation above the mean, such that 0.68 of the population are expected to have observations within one standard deviation of the mean. Similarly, 0.95 and 0.99 of the population are, respectively, expected to have observations within approximately 2 and 2.5 standard deviations of the mean. The actual number of standard deviations corresponding to the proportions of 0.95 and 0.99 are 1.960 and 2.576, respectively.

The expected proportion of a population with values greater than a given threshold point, measured in phenotypic standard deviations from the mean, is provided in the Appendix.

Given the properties of the normal distribution and knowledge of the mean and variance of a population, then properties of the distribution of observations for the population can be predicted. For example, if ultrasonic backfat depth has a mean of 15 mm and standard deviation of 3 mm in a particular population, then assuming a normal distribution, it is expected that 0.95 of the population would have backfat depths of between 9.1 and 20.9 mm, calculated from

$$15 - 1.96 \times 3 \quad \text{and} \quad 15 + 1.96 \times 3$$

Given several samples from the population, variation between the mean values of the samples is expected, just as there is variation between observations in a sample. If observations from the population have a normal distribution with mean μ and variance σ^2, then the mean values, estimated from samples of size N, will be normally distributed with mean μ, but with a variance of σ^2/N and a standard error of $\sqrt{\sigma^2/N}$.

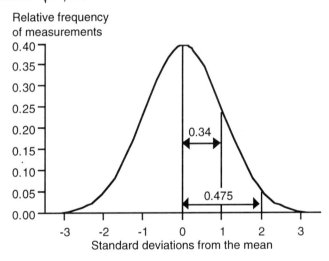

Fig. 1.1. Normal distribution

As the observations are normally distributed, then the sample means will have a normal distribution. As 0.95 of all observations are expected to be within two standard deviations of the population mean, then 0.95 of all sample means are expected to deviate from the population mean by less than two standard errors. If the population standard deviation is 3 mm and samples of size 100 are repeatedly taken, then it is expected that 0.95 of all sample means would deviate from the population mean by less than 0.6 mm, equal to $1.96\sqrt{3^2/100}$.

Analysis of Variance (ANOVA)

In animal breeding, prediction of the rate of genetic improvement requires information about the observed variation between animals and the contribution of genetic and environmental variation to the observed variation. The methodology of analysis of variance, ANOVA, is used to separate the total variation between observations into its component parts.

The ANOVA method is most easily explained with an example. Given s groups with n observations per group, the ANOVA method separates the total variation between the sn observations into the within-group variation and the between-group variation. The magnitude of the between-group variation is a measure of the size of differences between groups. If the group means are similar, then the between-group variation will be small.

The jth observation in the ith group is denoted X_{ij}, such that $\overline{X}_i$ is the mean of observations in group i and $\overline{X}$ is the overall mean. The deviation between an observation and the overall mean can be divided into two parts:
(1) the deviation between the observation and its group mean; and
(2) the deviation between the group mean and the overall mean

$$\left(X_{ij} - \overline{X}\right) = \left(X_{ij} - \overline{X}_i\right) + \left(\overline{X}_i - \overline{X}\right)$$

After squaring and summing terms, the equation leads to the total variance being separated into the within-group variance, the average variation between observations within a group, and the between-group variance, the variation between group means.

Observations can be laid out in a two-way table, with the n observations for each of the s groups in rows:

		\multicolumn{4}{c}{Observation}			
		1	2	...	n
	1	X_{11}	X_{12}	...	X_{1n}
Group	2	X_{21}	X_{22}	...	X_{2n}
	⋮				
	s	X_{s1}	X_{s2}	...	X_{sn}

The total sum of squares is calculated as

$$\sum_{i=1}^{s}\sum_{j=1}^{n}\left(X_{ij} - \overline{X}\right)^2 = \sum_{i=1}^{s}\sum_{j=1}^{n} X_{ij}^2 - \frac{1}{sn}\left(\sum_{i=1}^{s}\sum_{j=1}^{n} X_{ij}\right)^2$$

The term $\frac{1}{sn}\left(\sum_{i=1}^{s}\sum_{j=1}^{n} X_{ij}\right)^2$, equivalent to $sn\overline{X}^2$, is the correction factor, CF.

The total sum of squares is calculated as the sum of the squared observations minus the correction factor. Subtraction of the correction factor from the sum of squared observations enables the variation between observations to be expressed relative to the mean, rather than as deviations from zero. The total sum of squares has (sn−1) degrees of freedom, as there are sn observations in total.

The sum of squares for group means, when treated as "observations", is

$$\sum_{i=1}^{s} n(\overline{X}_i - \overline{X})^2 = \frac{1}{n} \sum_{i=1}^{s} (\text{group total})^2 - CF$$

The between-group sum of squares has (s−1) degrees of freedom, given s groups.
The within-group sum of squares can be calculated as

$$\sum_{i=1}^{s} \sum_{j=1}^{n} (X_{ij} - \overline{X}_i)^2$$

or as the difference between the total sum of squares and the between-group sum of squares. The within-group sum of squares has s(n−1) degrees of freedom, as there are s groups, each with (n−1) degrees of freedom.

The between-group and within-group mean squares, denoted by MS_B and MS_W respectively, are the sum of squares divided by their degrees of freedom, in a similar manner that the variance of N observations is the sum of squares divided by (N−1) degrees of freedom, as outlined in Table 1.2.

Table 1.2. Formula for calculating sums of squares and mean squares

Source of variation	DF	Sum of squares	Mean squares	
Between groups	s−1	$\frac{1}{n}\sum(\text{group total})^2 - CF$	Divide sum of squares by DF	MS_B
Within groups	s(n−1)	By difference		MS_W
Total	sn−1	$\sum X_{ij}^2 - CF$		

Balanced and unbalanced experimental designs

In a balanced design, each group has the same number of observations, while in an unbalanced design, all groups do not have the same number of observations. Calculation of the between-group sum of squares and the within-group degrees of freedom for the unbalanced design is slightly different from the balanced design.

The between-group sum of squares has to take account of the different number of observations in each group and is

$$\sum_{i=1}^{s} n_i (\overline{X}_i - \overline{X})^2 = \sum_{i=1}^{s} \frac{1}{n_i} (\text{group total})^2 - CF$$

The total sum of squares and the within-group sum of squares are calculated as for the balanced design. Similarly, the degrees of freedom for the between-group and the total sum of squares are s−1 and N−1, respectively, where N is the total number of observations. The degrees of freedom for the within-group sum of squares is N−s.

Expectation of Mean Squares

Assuming that the groups are random effects, then in a balanced design with n measurements per group, the between-group mean square is an unbiased estimate of which is equivalent to noting that the expectation of the between-group mean square is $\sigma_W^2 + n\sigma_B^2$, where σ_B^2 and σ_W^2 are the between-group and within-group variances, respectively.

For an unbalanced design, the expectation of the between-group mean square is $\sigma_W^2 + n_0\sigma_B^2$ with

$$n_0 = \frac{1}{s-1}\left[N - \frac{1}{N}\left(\sum_{i=1}^{s} n_i^2\right)\right]$$

for a total number of observations of N.

Estimates of the between-group and within-group variances are

$$\sigma_W^2 = MS_W$$

and

$$\sigma_B^2 = (MS_B - MS_W)/n$$

The variance of the estimated mean, for observations in a group, provides a measure of the precision with which the group mean is estimated. The variance of the estimated group mean is

$$\sigma_B^2 + \frac{\sigma_W^2}{n}$$

If the group mean was known without error, then the variance would be σ_B^2. Since the group mean is estimated from a sample of n observations, then the variance of the sampling error is $\frac{1}{n}\sigma_W^2$, where σ_W^2 is the variation within the group. Combining the two terms σ_B^2 and $\frac{1}{n}\sigma_W^2$, results in the variance of the estimated group mean.

Example
There are four groups with six observations per group:

Group	Observations					
A	9	11	10	9	9	12
B	9	10	8	8	7	6
C	5	8	6	9	8	9
D	12	16	13	15	14	17

The variance of measurements within groups A, B, C and D, calculated from the data, are 1.6, 2.0, 2.7 and 3.5. In particular, the mean and variance of group B are 8.0 and $\frac{1}{5}(9^2 + 10^2 + 8^2 + 8^2 + 7^2 + 6^2 - 6\times 8.0^2) = 2.0$. The average of the four within-group variances is $(1.6 + 2.0 + 2.7 + 3.5)/4 = 2.45$. Therefore, the estimate of the within-group variance, σ_W^2, is the average variance between measurements within groups.

The group means are 10.0, 8.0, 7.5 and 14.5, and the variance between the group means is

$$\frac{1}{3}\left(10.0^2 + 8.0^2 + 7.5^2 + 14.5^2 - 4 \times 10^2\right) = 10.17$$

The between-group mean square divided by the number of observations per group is $61.0/6 = 10.17$. Therefore, the variance of a group mean, $\sigma_B^2 + \frac{\sigma_W^2}{n}$, is estimated from the variance of the group means.

The analysis of variance table is as follows:

Source of variation	DF	Mean squares	Expectation of mean squares
Between groups	3	61.00	$\sigma_W^2 + 6\sigma_B^2$
Within groups	20	2.45	σ_W^2

Example data set

The following data set will be used on several occasions throughout this chapter and Chapter 2, to demonstrate how information on variances is used in a quantitative genetics framework. In each of eight groups, there are three observations:

Group	Observations			Group total
1	28	29	27	84
2	30	33	31	94
3	28	28	25	81
4	30	31	28	29
5	26	26	28	80
6	27	29	29	85
7	27	28	27	82
8	25	21	32	78

First, the correction factor, CF, is calculated, which is the square of the sum of the observations, 673, divided by the number of observations, 24. Therefore, CF = $673^2/24$ = 18872.04.

The total sum of squares is the sum of the observations squared minus the correction factor, such that the total sum of squares is

$$(28^2 + 29^2 + 27^2 + 30^2 + \ldots + 21^2 + 32^2) - CF = 148.96$$

The between-group sum of squares is the sum of the group totals squared, divided by the number of observations per group, and then the correction factor is subtracted. The between group-sum of squares is

$$(84^2 + 94^2 + 81^2 + \ldots + 82^2 + 78^2)/3 - CF = 63.63$$

Source of variation	DF	Sum of squares	Mean squares	Expectation of Mean squares
Between groups	7	63.63	9.09	$\sigma_W^2 + 3\sigma_B^2$
Within groups	16	85.33	5.33	σ_W^2
Total	23	148.96		

The between-group and within-group variances are estimated by equating the mean squares to their expectations. Estimates of the within-group and between-group variances, σ_W^2 and σ_B^2, are 5.33 and (9.09 − 5.33)/3 = 1.25, respectively.

Repeatability

Now that the between-group, σ_B^2, and within-group, σ_W^2, variances have been estimated, what can be done with them? One parameter of interest is the repeatability, r_e, which is a measure of the similarity between the observations within a group. The repeatability is

$$r_e = \frac{\sigma_B^2}{\sigma_B^2 + \sigma_W^2}$$

The repeatability is also known as the intra-class correlation, which is the expected correlation between observations within the group, or class. The correlation between observations is discussed in Chapter 3.

The ratio of the between-group variance to the sum of the between-group and within-group variances is a measure of the relative contribution of the within-group variance to the total variation. If σ_W^2 is small, then observations within a group are similar and highly repeatable. Conversely, a large value for σ_W^2 indicates that there is substantial variation between observations within a group, and so the repeatability will be low.

A formula for the variance of the repeatability estimate is

$$\text{var}(r_e) = \frac{2(1-r_e)^2 (1+(n-1)r_e)^2}{(s-1)n(n-1)}$$

where s and n are the number of groups and the number of observations per group, respectively (Robertson, 1959b). The variance of the repeatability estimate provides an indication of the precision with which the repeatability was estimated.

In the previous example, estimates of σ_B^2 and σ_W^2 are 1.25 and 5.33, respectively, such that the estimate of the repeatability is

$$r_e = \frac{1.25}{1.25 + 5.33} = 0.19$$

The variance of an estimated group mean was previously shown to be

$$\sigma_B^2 + \frac{\sigma_W^2}{n}$$

but when the variance components, σ_B^2 and σ_W^2, are expressed in terms of the repeatability:

$$\sigma_B^2 = r_e\left(\sigma_B^2 + \sigma_W^2\right) \quad \text{and} \quad \sigma_W^2 = (1-r_e)\left(\sigma_B^2 + \sigma_W^2\right)$$

then the variance of an estimated group mean is

$$\left[r_e + \frac{1-r_e}{n}\right]\left(\sigma_B^2 + \sigma_W^2\right)$$

or

$$\left[\frac{1+(n-1)r_e}{n}\right]\left(\sigma_B^2 + \sigma_W^2\right)$$

Chapter two

ANOVA in a Quantitative Genetics Framework

In a quantitative genetics framework, the ANOVA method is used to separate the total variation between observations into its components, such as the genetic and environmental variation. The heritability is a function of the genetic and environmental variances and is central to animal breeding theory. Both the predicted rate of genetic improvement in a population and the prediction of an animal's genetic merit depend on the heritability.

Animals can be grouped according to their sire, such that the total variation between animals can be separated into variation between animals of different sire-families and the variation within sire-families. Dairy cows are an example of grouping animals according to their sire, as there are often large half-sib families. The variation between sire-families and within sire-families provides information on the heritability. If the observations are grouped according to their dam, the variation between dam-families and within dam-families can provide information on maternal and common environmental effects. Pigs can be categorised into full-sib groups, according to their dam, but if pigs are categorised according to their sire, then sire-families will consist of both full-sibs and half-sibs.

Variance components estimated with ANOVA are used to provide estimates of the heritability and the common environmental effect, as both parameters are required for prediction of genetic merit. The most straightforward method of explaining the use of ANOVA to separate the total variation between animals into the different variance components is with an example. Assume that there are s sires, each mated to d dams, and each dam has n progeny, such that there is a total of sdn animals. As in Chapter 1, the correction factor, CF, is calculated, which is the square of the sum of the observations, divided by the number of observations

$$CF = (\Sigma X)^2 / sdn$$

where X represents an observation.

The total sum of squares, Total SS, is the sum of the squares of the observations minus the correction factor:

$$\text{Total SS} = \Sigma X^2 - CF$$

The between-sire sum of squares, Sire SS, is the sum of the squared sire totals, divided by the number of progeny per sire, minus the correction factor:

$$\text{Sire SS} = \frac{1}{dn}\sum(\text{sire total})^2 - \text{CF}$$

The sum of squares between-dams, but within-sires, Dam SS, is the sum of the squared dam totals, divided by the number of progeny per dam, minus both the correction factor and the sire sum of squares:

$$\text{Dam SS} = \frac{1}{n}\sum(\text{dam total})^2 - \text{CF} - \text{Sire SS}$$

The within-dam sum of squares is the total sum of squares minus both the between-sire and the between-dam within-sire sum of squares.

If the number of animals per dam or per sire is not the same for each dam, or sire, an unbalanced design, then the between-sire sum of squares is

$$\sum\frac{(\text{sire total})^2}{n_i} - \text{CF}$$

where n_i is the number of progeny for the ith sire and the between-dam within-sire sum of squares is

$$\sum\frac{(\text{dam total})^2}{n_{ij}} - \text{CF} - \text{Sire SS}$$

where n_{ij} is the number of progeny of the jth dam mated to the ith sire.

However, the equation for calculating the correction factor and the total sum of squares is the same for both balanced and unbalanced designs.

The analysis of variance table, for a balanced design, is shown in Table 2.1.

Table 2.1. Formula for calculating sums of squares and mean squares

Source of variation	DF	Sum of squares	Expectation of mean squares
Between sires	s−1	$\frac{1}{dn}\sum(\text{sire total})^2 - \text{CF}$	$\sigma_e^2 + n\sigma_d^2 + nd\sigma_s^2$
Within-sires and between dams	s(d−1)	$\frac{1}{n}\sum(\text{dam total})^2 - \text{CF} - \text{Sire SS}$	$\sigma_e^2 + n\sigma_d^2$
Within-dams	sd(n−1)	By difference	σ_e^2
Total	sdn−1	$\sum X^2 - \text{CF}$	

Variance Components

The between-sires, between-dams and within-dams, or residual, mean squares are denoted by MS_s, MS_d and MS_e, respectively. When the mean squares are equated to their expectations, the estimated variance components are as follows:

between-sire variance $\quad \sigma_s^2 = (MS_s - MS_d)/nd$
between-dam variance $\quad \sigma_d^2 = (MS_d - MS_e)/n$
residual variance $\quad \sigma_e^2 = MS_e$

The estimated variance components are used to calculate functions of the variance components, such as

phenotypic variance $\quad \sigma_P^2 = \sigma_e^2 + \sigma_d^2 + \sigma_s^2$
half-sib correlation $\quad t_{HS} = \sigma_s^2/\sigma_P^2$
full-sib correlation $\quad t_{FS} = (\sigma_s^2 + \sigma_d^2)/\sigma_P^2$

The term phenotypic variance is discussed later in this chapter. Estimates of the half-sib and full-sib correlations provide an indication of the repeatability of measurements between half-sibs and between full-sibs, respectively.

Example

The previous data set is used, but now the data are grouped according to sires and dams. Dams 1 and 2 are mated to sire 1, dams 3 and 4 are mated to sire 2, dams 5 and 6 are mated to sire 3, and dams 7 and 8 are mated to sire 4:

Dam	Observations			Dam total	Sire	Sire total
1	28	29	27	84	1	178
2	30	33	31	94	1	
3	28	28	25	81	2	170
4	30	31	28	29	2	
5	26	26	28	80	3	165
6	27	29	29	85	3	
7	27	28	27	82	4	160
8	25	21	32	78	4	

The dam sum of squares is calculated as:

$$(84^2 + 94^2 + \ldots + 82^2 + 78^2)/3 - CF - \text{Sire SS} = 34.17$$

Source of variation	DF	Sum of squares	Mean squares	Expectation of mean squares
Between-sires	3	29.46	9.82	$\sigma_e^2 + 3\sigma_d^2 + 6\sigma_s^2$
Within-sires and between-dams	4	34.17	8.54	$\sigma_e^2 + 3\sigma_d^2$
Within-dams	16	85.33	5.33	σ_e^2
Total	23	148.96		

The sire, σ_s^2, dam, σ_d^2, and residual variances, σ_e^2, are estimated by equating mean squares to their expectations. The estimated variance components are 0.21, 1.07 and 5.33, respectively.

The estimated phenotypic variance, $\sigma_P^2 = \sigma_e^2 + \sigma_d^2 + \sigma_s^2$, is 6.61.
Estimates of the half-sib, $t_{HS} = \sigma_s^2/\sigma_P^2$, and full-sib, $t_{FS} = \left(\sigma_s^2 + \sigma_d^2\right)/\sigma_P^2$, correlations are 0.03 and 0.19, respectively.

The variance of a sire-family mean is

$$\sigma_s^2 + \sigma_d^2/d + \sigma_e^2/nd \quad \text{equal to} \quad MS_s/nd = 1.64$$

The variance of a dam-family mean is

$$\sigma_d^2 + \sigma_e^2/n + \sigma_s^2 \quad \text{equal to} \quad MS_d/n + \sigma_s^2 = 3.06$$

The variance of a dam family mean includes the sire variance component, as variation between sires will be reflected in variation between dams, due to the hierarchical mating structure.

Precision of Estimated Parameters

Precision of the estimated parameters, such as the full-sib and half-sib correlations, is inversely related to the variance of the estimated parameters. As the estimated parameters are derived from the mean squares, and then the variance of the mean squares is used to determine the variance of the estimated parameters.

The estimated variance of a mean square is twice the square of the mean square divided by its degrees of freedom plus two:

$$\text{var}(MS) = \frac{2MS^2}{DF + 2}$$

and the variance has a χ^2 distribution.

The dam variance component is a function of the dam and residual mean squares, $(MS_d - MS_e)/n$, so the estimated variance of the estimated dam variance component is

$$\text{var}(\sigma_d^2) = \frac{1}{n^2}\text{var}(MS_d) + \frac{1}{n^2}\text{var}(MS_e)$$

$$= \frac{2}{n^2}\left[\frac{MS_d^2}{s(d-1)+2} + \frac{MS_e^2}{sd(n-1)+2}\right]$$

Similarly, the estimated variance of the estimated sire variance component is

$$\text{var}(\sigma_s^2) = \frac{2}{n^2 d^2}\left[\frac{MS_s^2}{(s-1)+2} + \frac{MS_d^2}{s(d-1)+2}\right]$$

When measurements are taken on half-sibs only, such that dams are not included in the model, then the between-sire mean square is an estimate of $\sigma_e^2 + n\sigma_s^2$ and the estimated variance of the estimated sire variance component is

$$\text{var}(\sigma_s^2) = \frac{2}{n^2}\left[\frac{MS_s^2}{(s-1)+2} + \frac{MS_e^2}{s(n-1)+2}\right]$$

The variance of the estimated half-sib correlation, with one progeny per dam, is

$$\text{var}(t_{HS}) = \frac{2(1-t_{HS})^2(1+(n-1)t_{HS})^2}{(s-1)n(n-1)}$$

When the number of progeny per sire, n, is approximately $1/t_{HS}$, the variance of the estimated half-sib correlation can be approximated (Robertson, 1959a) as

$$\text{var}(t_{HS}) \approx \frac{8(1-t_{HS})^2 t_{HS}}{sn}$$

Effect of including only sires or dams in the model

If only sires were included in the model for analysis of the example data, then the between-sire mean square would remain the same as when both sires and dams were included in the model, but the residual mean square would increase to $(34.17+85.33)/20 = 5.97$. The expectation of the sire mean square would estimate $\sigma_e^2 + 6\sigma_s^2$, and the estimated sire variance would be 0.64. The estimated phenotypic variance would remain unchanged, at 6.61, but the half-sib correlation would increase from 0.03 to 0.10.

If only dams were included in the model, then the dam mean square would change to $(29.46+34.17)/7 = 9.09$. The expectation of the dam mean square would still be $\sigma_e^2 + 3\sigma_d^2$, but the estimated dam variance component would increase to 1.25, with the estimated phenotypic variance decreasing to 6.58 (see Table 2.2).

Table 2.2. Estimates of variance components assuming different models

Model	Variance components				Relationship between progeny
	σ_s^2	σ_d^2	σ_e^2	σ_P^2	
Sires and dams	0.21	1.07	5.33	6.61	Full and half-sibs
Sires only	0.64	—	5.97	6.61	Half-sibs
Dams only	—	1.25	5.33	6.58	Full-sibs

The effect of including only sires in the model is that estimates of both the sire and residual variances are inflated to account for the variation between groups of full-sibs from the same sire. The sire and residual variances are increased by

$$\frac{(n-1)}{(dn-1)}\sigma_d^2 \quad \text{and} \quad \frac{(d-1)n}{(dn-1)}\sigma_d^2, \text{ respectively,}$$

but the estimate of the phenotypic variance is unchanged. If d equals one, there will be no bias in the residual variance, but the estimated sire variance would be increased by the dam variance, as all progeny of a sire would be full-sibs. Conversely, if n equals one, then the estimated sire variance is unbiased, as all progeny of a sire will be half-sibs, but the estimated residual variance is increased by the dam variance.

If only dams are included in the model, then the estimated dam variance is increased by $\frac{d(s-1)}{(sd-1)}\sigma_s^2$, but the estimated phenotypic variance is reduced by $\frac{(d-1)}{(sd-1)}\sigma_s^2$. The sum of the changes in the estimated dam and phenotypic variances is the sire variance component, such that the estimate of the residual variance is unchanged.

Clearly, the choice of model has an effect on the estimated variances. Therefore, it is important that the chosen model takes account of the genetic structure of the data to ensure appropriate estimates of the genetic parameters.

Genetic Interpretation of Variance Components

One model assumes that the animal's measurement, its phenotype, P, consists of a genetic, G, and environmental, E, effect. The genotype can be divided into an additive genetic effect, A, a maternal genetic effect, M, and a dominance effect, D. The environmental effect can be separated into a common environmental effect, E_C, and the general environmental effect, E. The model $P = G + E$ can be extended to

$$P = \underbrace{A + M + D}_{\text{genetic}} + \underbrace{E_C + E}_{\text{environmental}}$$

For further information, consult Falconer and MacKay (1996).

The additive genetic effect is the sum of the average effects of genes, with the summation over each pair of alleles at each locus, for all loci. The maternal genetic effect is the influence of the mother's phenotype on the phenotype of her progeny, which increases the resemblance between the dam and her progeny. For example, larger dams may give more milk to their progeny, which are larger than offspring from smaller dams. The dominance effect is due to dominance between alleles at a locus, or to the effect not accounted for by the average effects of genes.

The common environmental effect increases the similarity between full-sibs, as they share the same environment. The common environmental effect differs from the maternal genetic effect, as the maternal genetic effect directly increases the similarity between offspring and dam, so indirectly increasing the similarity between progeny, while the common environmental effect increases the similarity between progeny directly. The general environmental effect is a result of differences between animals in nutrition, management etc., but is not due to factors relating to the animal's genotype.

Relationship of variance components to genetic and environmental parameters

Estimates of the sire, dam and residual variance components are used to quantify the different genetic and environmental contributions to the phenotype (Thompson, 1976).

The sire variance component is a quarter of the additive genetic variance, which is the variance of the additive genetic effect. The dam variance component also includes a quarter of the additive genetic variance, but the maternal genetic effect, a quarter of the dominance effect and the common environmental effect all contribute to the dam variance component (see Table 2.3). Therefore, certain assumptions must be made about the composition of the dam variance component, as there are too many parameters to estimate simultaneously. For example, dominance may be ignored. Estimation of the maternal genetic effect requires information on progeny from several litters of each dam, such that the common environmental effect can be separated from the maternal genetic effect.

Table 2.3. Expectation of sire, dam and residual variance components

Variance	Genetic contribution			Environmental contribution	
	Additive (A)	Maternal (M)	Dominance (D)	Common (E_C)	General (E)
Sire	1/4				
Dam	1/4	1	1/4	1	
Residual	1/2		3/4		1

If maternal genetic and common environmental effects are combined and dominance variance is ignored, then the phenotypic variance is the sum of the additive genetic, maternal/common environmental and environmental variances:

$$\sigma_P^2 = \sigma_A^2 + \sigma_M^2 + \sigma_E^2$$

Equating the sire, dam and residual variance components to their expectations in terms of the additive genetic variance, the maternal variance and the environmental variance provides the estimates

$$\sigma_s^2 = \frac{1}{4}\sigma_A^2 \qquad\qquad \sigma_A^2 = 4\sigma_s^2$$

$$\sigma_d^2 = \frac{1}{4}\sigma_A^2 + \sigma_M^2 \qquad\qquad \sigma_M^2 = \sigma_d^2 - \sigma_s^2$$

$$\sigma_e^2 = \sigma_P^2 - \frac{1}{2}\sigma_A^2 - \sigma_M^2 = \sigma_E^2 + \frac{1}{2}\sigma_A^2 \qquad \sigma_E^2 = \sigma_e^2 - 2\sigma_s^2$$

When all progeny are half-sibs, then the dam is not included in the model, and the estimated sire and residual variance components can be equated to the additive genetic variance and the environmental variance:

$$\sigma_s^2 = \frac{1}{4}\sigma_A^2 \qquad\qquad \sigma_A^2 = 4\sigma_s^2$$

$$\sigma_e^2 = \sigma_P^2 - \frac{1}{4}\sigma_A^2 = \sigma_E^2 + \frac{3}{4}\sigma_A^2 \qquad \sigma_E^2 = \sigma_e^2 - 3\sigma_s^2$$

Lower case subscripts are used for the sire, dam and residual variance components and their corresponding mean squares, while upper case subscripts are used for the additive genetic variance, the maternal variance and the environmental variance. The reason for the difference in case is to prevent confusion between the residual variance component and the environmental variance.

Heritability

The heritability is defined as the proportion of the phenotypic variance attributed to the additive genetic variance. A high heritability indicates that a substantial proportion of the phenotypic variance is due to additive genetic variation, while non-additive genetic factors make a larger contribution to the phenotype when the heritability is low. The heritability is

$$h^2 = \frac{\sigma_A^2}{\sigma_P^2}$$

where σ_A^2 is the additive genetic variance. Examples of heritability estimates are 0.09 for litter size in pigs (Haley and Lee, 1992), 0.39 for milk yield in dairy cows (Swalve, 1995) and 0.63 for carcass lean content in pigs (Hovenier *et al.*, 1992).

The heritability can be estimated from estimates of the additive genetic variance and phenotypic variance, or from estimates of the half-sib correlation or the full-sib correlation. Since the expectation of the sire variance component is a quarter of the additive genetic variance, then

$$h^2 = 4\frac{\sigma_s^2}{\sigma_P^2} = 4t_{HS}$$

The variance of the estimated heritability is

$$\text{var}(h^2) = \frac{32(1-t_{HS})^2(1+(n-1)t_{HS})^2}{(s-1)n(n-1)}$$

$$= 16\,\text{var}(t_{HS}) \approx 16\frac{\text{var}(\sigma_s^2)}{\sigma_P^4}$$

A formula for the variance of the estimated intra-class correlation or repeatability was given in Chapter 1.

If the heritability is estimated from the full-sib correlation, then the heritability estimate may be biased, as the maternal effect, the common environmental effect and the dominance effect are included in the dam variance component.

In the example data set, the half-sib and full-sib correlations were 0.03 and 0.16, respectively, such that the corresponding heritability estimates were 0.13 and 0.65.

Maternal Effect

The combination of the maternal and common environmental effects and the dominance effect, denoted by c^2, can be estimated from the half-sib and full-sib correlations, as

$$c^2 = t_{FS} - t_{HS}$$

An approximate variance of the maternal and common environmental effect is

$$\operatorname{var}(c^2) \approx \frac{\operatorname{var}(\sigma_d^2 - \sigma_s^2)}{\sigma_P^4} = \frac{\operatorname{var}(\sigma_d^2)}{\sigma_P^4} + \frac{\operatorname{var}(\sigma_s^2)}{\sigma_P^4}$$

assuming that the sire and dam effects are independent.

Examples of maternal/common environmental effects in pigs for the performance test traits of growth rate, ultrasonic backfat depth and food conversion ratio are 0.14, 0.05 and 0.09, with corresponding heritability estimates of 0.30, 0.64 and 0.22 (Ducos *et al.*, 1993).

Chapter three
Regression and Correlation

So far, only one trait has been considered, but in many situations the association between traits is of interest. For example, if a particular breeding programme which increases carcass lean content also reduces reproductive performance, then prior information about the negative association between the two traits could have been used to prevent, or at least constrain, the reduction in reproductive performance. Secondly, measurement of the relationship between parent and offspring can provide information about genetic parameters, such as the heritability.

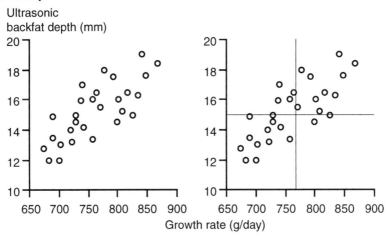

Fig. 3.1. Growth rate and ultrasonic backfat depth of pigs

If the two traits are plotted with vertical and horizontal lines drawn at the mean value of each trait, then the graph can be divided into four areas (Fig. 3.1). If most of the data points lie in the bottom-left and top-right quadrants, then the trait on the Y-axis increases as the trait on the X-axis increases and the traits are positively related. Conversely, if the majority of data points lie in the top-left and bottom-right quadrants, then the trait on the Y-axis decreases as the trait on the X-axis increases and the traits are negatively related:

$(X-\overline{X}) < 0$	$(X-\overline{X}) > 0$
$(Y-\overline{Y}) > 0$	$(Y-\overline{Y}) > 0$
$(X-\overline{X}) < 0$	$(X-\overline{X}) > 0$
$(Y-\overline{Y}) < 0$	$(Y-\overline{Y}) < 0$

If the traits are positively related, then the product of $(X - \overline{X})$ and $(Y - \overline{Y})$ will generally be positive, as data points in the top-right quadrant will both be greater than their respective means and data points in the bottom-left quadrant will both be less than their respective means. In contrast, if the traits are negatively related, then the product of $(X - \overline{X})$ and $(Y - \overline{Y})$, in both the top-left and bottom-right quadrants, will generally be negative.

Covariance

The covariance between traits X and Y is denoted by cov(X,Y) or σ_{XY}, just as the variance of X is denoted by var(X) or σ_X^2.

The covariance is estimated by

$$\text{cov}(X, Y) = \frac{1}{N-1} \sum_{i=1}^{N} (X_i - \overline{X})(Y_i - \overline{Y})$$

where N is the number of X and Y pairs in the sample. The equation for the covariance is of the same form as that for the variance:

$$\text{var}(X) = \frac{1}{N-1} \sum_{i=1}^{N} (X_i - \overline{X})^2$$

Just as the variance was calculated as

$$\frac{1}{N-1} \left[\sum_{i=1}^{N} X_i^2 - N\overline{X}^2 \right]$$

then rather than squaring deviations from the mean, the covariance can be calculated as

$$\text{cov}(X, Y) = \frac{1}{N-1} \left[\sum_{i=1}^{N} X_i Y_i - N\overline{X}\overline{Y} \right]$$

If a linear relationship between traits X and Y is assumed, such that, for every unit increase in X, there is a corresponding change in Y of b_{YX} units (Fig. 3.2), then

$$Y = a + b_{YX} X$$

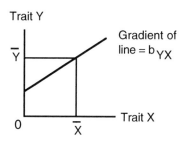

Fig. 3.2. Linear regression of trait Y on trait X

The estimated regression coefficient of trait Y on trait X, b_{YX}, is the covariance between X and Y divided by the variance of X

$$b_{YX} = \frac{\text{cov}(X, Y)}{\text{var}(X)}$$

with the regression coefficient subscripts denoting that it is the regression of Y on X. The regression line passes through the means of both traits, such that the estimate of the intercept, a, is

$$a = \overline{Y} - b_{YX}\overline{X}$$

The regression equation can be written as

$$Y = (\overline{Y} - b_{YX}\overline{X}) + b_{YX}X \quad \text{or as} \quad (Y - \overline{Y}) = b_{YX}(X - \overline{X})$$

The estimated regression coefficient is such that the sum of the squared deviations between the observed values of Y and the regression line are minimised. For further information see Snedecor and Cochran (1989).

One use of the estimated regression equation is to predict Y for a given value of X, particularly in an animal breeding context, when an animal's additive genetic effect is predicted from its phenotype.

Properties of the variance

The condition of independence, which was assumed in Chapter 1, can now be dropped to take account of the linear relationship between traits X and Y, such that the fifth property of the variance is as follows:

Property 5 $\quad \text{var}(cX + dY) = c^2 \text{var}(X) + d^2 \text{var}(Y) + 2cd\,\text{cov}(X,Y)$
where c and d are constants.

If the constants c and d are equal to one, then:

$$\text{var}(X + Y) = \text{var}(X) + \text{var}(Y) + 2\,\text{cov}(X, Y)$$

and when constants c and d are equal to plus and minus one, then:

$$\text{var}(X - Y) = \text{var}(X) + \text{var}(Y) - 2\,\text{cov}(X, Y)$$

Therefore, $\text{var}(X + Y) \neq \text{var}(X - Y)$ when the regression coefficient, b_{YX}, or the covariance between X and Y, is not zero.

Correlation between Traits

The regression coefficient describes the linear relationship between two traits, while the correlation coefficient is a measure of the variation in trait Y attributable to the linear relationship with trait X.

The two distributions in Fig. 3.3 have the same regression coefficient and intercept, but there is considerably more variation about the regression line in one distribution compared to the other.

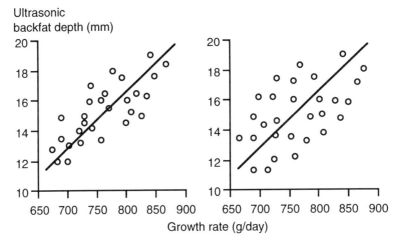

Fig. 3.3. Growth rate and ultrasonic backfat depth for two groups of pigs

The correlation coefficient is

$$r = \frac{\text{cov}(X, Y)}{\sqrt{\text{var}(X)\text{var}(Y)}} = \frac{\sigma_{XY}}{\sigma_X \sigma_Y}$$

The correlation coefficient lies between plus and minus one. The regression, b_{YX}, and correlation, r, coefficients can be derived from each other:

$$b_{YX} = \frac{\sigma_{XY}}{\sigma_X^2} = \frac{\sigma_{XY}}{\sigma_X \sigma_Y} \frac{\sigma_Y}{\sigma_X} = \frac{\sigma_Y}{\sigma_X} r$$

The estimate of the correlation coefficient is not normally distributed. However, the parameter

$$z = 0.5 \ln\left(\frac{1+r}{1-r}\right)$$

is almost normally distributed with an approximate standard error of $\sqrt{1/(N-3)}$, for N pairs of observations on X and Y. Back-transformation of the standard error of z will provide an estimate of the standard error of the correlation coefficient, using the transformation

$$\text{s.e.}(r) = \frac{e^{2\text{se}(z)} - 1}{e^{2\text{se}(z)} + 1}$$

Analysis of variance for linear regression

Analysis of variance is used to examine the linear relationship between traits X and Y. The analysis of variance table is as shown in Table 3.1.

Table 3.1. Formula for calculating analysis of variance table

Source of variation	DF	Sum of squares	$\dfrac{\text{Sum of squares}}{N-1}$
Regression	1	$b_{YX}\left[\sum XY - \dfrac{\sum X \sum Y}{N}\right]$	$r^2 \sigma_Y^2$
Residual	N–2	By difference	$(1-r^2)\sigma_Y^2$
Total	N–1	$\sum(Y-\overline{Y})^2$	σ_Y^2

The regression sum of squares can be expressed as a function of the correlation coefficient, since

$$b_{YX}(N-1)\sigma_{XY} = \dfrac{\sigma_{XY}^2}{\sigma_X^2}(N-1) = \dfrac{\sigma_{XY}^2}{\sigma_X^2 \sigma_Y^2}\sigma_Y^2(N-1) = r^2\sigma_Y^2(N-1)$$

The proportion of variance in trait Y, not accounted for by the linear regression, is essentially equal to $(1-r^2)$, as the residual mean square is

$$(1-r^2)\sigma_Y^2 \dfrac{(N-1)}{(N-2)}$$

The standard error of the regression coefficient reflects the variation in the dependent variable, Y, that is accounted for by the linear regression on the independent variable, X. The variance of the estimated regression coefficient is the residual mean square divided by the sums of squares of the independent variable, such that

$$\text{var}(b_{YX}) = \dfrac{(1-r^2)}{N-2}\dfrac{\sigma_Y^2}{\sigma_X^2}$$

The estimated regression coefficient has a t distribution with (N–2) degrees of freedom.

Although the precision of the regression coefficient is determined from the standard error of the z parameter, where $z = 0.5\ln\left(\dfrac{1+r}{1-r}\right)$, an approximation to the variance of the correlation coefficient is

$$\text{var}(r) \approx \dfrac{1-r^2}{N-2}$$

as $\quad r = \dfrac{\sigma_X}{\sigma_Y}b_{YX}$

Example
Calculate the regression equation and the correlation coefficient for offspring weight on parental mean weight, using the following data:

Mean of parents	18.0	21.6	20.0	18.0	24.4	17.6	19.6	24.0	20.0	18.8
Offspring	18.4	20.4	19.2	17.6	22.0	18.8	20.4	20.8	19.6	20.4

A plot of the data (Fig. 3.4) reveals that there is a linear relationship between parental mean weight and offspring weight.

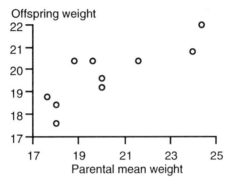

Fig. 3.4. Parental mean weight and offspring weight

To calculate the regression equation and the correlation coefficient, the following totals are required, with parental mean weight and offspring weight denoted by X and Y, respectively:

$\Sigma X = 202.0$ $\Sigma X^2 = 4133.28$ $\Sigma XY = 4014.88$
$\Sigma Y = 197.6$ $\Sigma Y^2 = 3919.68$

The variance of X $\sigma_X^2 = \dfrac{1}{N-1}\left[\Sigma X^2 - \dfrac{(\Sigma X)^2}{N}\right]$ = 5.876

The covariance $\sigma_{XY} = \dfrac{1}{N-1}\left[\Sigma XY - \dfrac{\Sigma X \Sigma Y}{N}\right]$ = 2.596

The regression coefficient $b_{YX} = \dfrac{\sigma_{XY}}{\sigma_X^2}$ = 0.442

The intercept $a = \overline{Y} - b_{YX}\overline{X}$ = 10.837

The correlation coefficient $r = \dfrac{\sigma_{XY}}{\sigma_X \sigma_Y}$ = 0.827

The regression equation is

offspring weight = 10.83 + 0.44 average parental weight

The analysis of variance table, for the example, is as follows:

Source of variation	DF	Sum of squares
Regression	1	10.32
Residual	8	4.78
Total	9	15.10

The regression coefficient is 0.44, with a standard error of 0.11, and the correlation coefficient is 0.83, with an approximate standard error of 0.20.

Genetic Interpretation of Regression and Correlation Coefficients

The regression coefficient of offspring measurements on parental measurements can be interpreted in a genetic framework, assuming that phenotype, P, is the sum of additive genetic, G, and environmental, E, effects:
$$P = G + E$$
The covariance between offspring phenotype and the phenotype of one parent is

$$\text{cov}(P_O, P_P) = \text{cov}(G_O + E_O, G_P + E_P)$$
$$= \text{cov}(G_O, G_P) + \text{cov}(G_O, E_P) + \text{cov}(E_O, G_P) + \text{cov}(E_O, E_P)$$

where the subscripts O and P denote the offspring and parent. Assuming that there is no covariance between genotype and environment, or between the environment of the offspring and the environment of the parent, then
$$\text{cov}(P_O, P_P) = \text{cov}(G_O, G_P)$$
Assuming that $P = G + E$, then the genotype of the offspring is one half of the additive genetic merit of one parent, such that $\text{cov}(G_O, G_P)$ is the covariance between G and $\frac{1}{2}G$. Since $\text{var}(G) = \sigma_A^2$, then
$$\text{cov}(P_O, P_P) = \text{cov}(G_O, G_P) = \frac{1}{2}\sigma_A^2$$
The regression coefficient of offspring phenotype on the phenotype of one parent is as follows:
$$b_{OP} = \frac{\text{cov}(P_O, P_P)}{\text{var}(P_P)} = \frac{\frac{1}{2}\sigma_A^2}{\sigma_P^2} = \frac{1}{2}h^2$$
as the variance of the parent's phenotype is the phenotypic variance, assuming the parent is a random sample from the population. The regression coefficient is also equal to $\frac{1}{2}h^2$ if the parents have been specifically selected for the purpose of estimating the regression of offspring on parent (Hill and Thompson, 1977).

The correlation between offspring phenotype and the phenotype of one parent is as follows:
$$r_{OP} = \frac{\text{cov}(P_O, P_P)}{\sqrt{\text{var}(P_O)\text{var}(P_P)}} = \frac{\frac{1}{2}\sigma_A^2}{\sqrt{\sigma_P^2 \sigma_P^2}} = \frac{1}{2}\frac{\sigma_A^2}{\sigma_P^2} = \frac{1}{2}h^2$$

When offspring phenotype is regressed on the mean phenotype of both parents, the covariance is $\frac{1}{2}\sigma_A^2$, since

$$\text{cov}\left[P_O, \frac{1}{2}(P_{P1}+P_{P2})\right] = \frac{1}{2}\left[\text{cov}(P_O, P_{P1}) + \text{cov}(P_O, P_{P2})\right] = \text{cov}(P_O, P_P)$$

where P_{P1} and P_{P2} are the phenotypes of the parents.

The variance of parental mean phenotype is $\sigma_P^2/2$, assuming that there is no covariance between the parental phenotypes, just as the variance of a mean of n independent observations is σ^2/n.

The regression of offspring phenotype on parental mean phenotype is

$$b_{O\bar{P}} = \frac{\text{cov}(P_O, P_{\bar{P}})}{\text{var}(P_{\bar{P}})} = \frac{\frac{1}{2}\sigma_A^2}{\frac{1}{2}\sigma_P^2} = \frac{\sigma_A^2}{\sigma_P^2} = h^2$$

and the correlation coefficient is

$$r_{O\bar{P}} = \frac{\text{cov}(P_O, P_{\bar{P}})}{\sqrt{\text{var}(P_O)\text{var}(P_{\bar{P}})}} = \frac{\frac{1}{2}\sigma_A^2}{\sqrt{\sigma_P^2 \frac{1}{2}\sigma_P^2}} = \sqrt{\frac{1}{2}\frac{\sigma_A^2}{\sigma_P^2}} = \sqrt{\frac{1}{2}h^2}$$

The heritability can be estimated from the regression of offspring phenotype on the phenotype of one parent or on parental mean phenotype.

Genetic Correlation between Traits

The covariance between two traits can be estimated at both the phenotypic and genetic levels, in a manner similar to estimation of the phenotypic and genetic variance of one trait.

The sums of squares and mean squares are replaced with sums of cross-products and mean cross-products between traits X and Y. For example, with s sires each with n half-sib progeny, the analysis of covariance table is as shown in Table 3.2.

Table 3.2. Formula for sums of cross-products and mean cross-products

Source of covariation	DF	Sum of cross-products	Expectation of mean cross-products
Between sires	s−1	$\frac{1}{n}\Sigma(\text{sire total}_X)(\text{sire total}_Y) - CF$	$\sigma_{e(XY)} + n\sigma_{s(XY)}$
Within sires	s(n−1)	By difference	$\sigma_{e(XY)}$
Total	sn−1	$\Sigma XY - CF$	

The correction factor, CF, of $N\overline{XY}$ and the mean cross-products is expressed in terms of residual and sire covariance components, $\sigma_{e(XY)}$ and $\sigma_{s(XY)}$.

Regression and correlation

The estimated sire and residual covariance components are equated to the additive genetic, $\sigma_{A(XY)}$, environmental, $\sigma_{E(XY)}$, and phenotypic, $\sigma_{P(XY)}$, covariances, as follows:

$$\sigma_{s(XY)} = \frac{1}{4}\sigma_{A(XY)}$$

$$\sigma_{e(XY)} = \sigma_{P(XY)} - \frac{1}{4}\sigma_{A(XY)} \quad \text{or} \quad \begin{aligned} \sigma_{A(XY)} &= 4\sigma_{s(XY)} \\ \sigma_{E(XY)} &= \sigma_{e(XY)} - 3\sigma_{s(XY)} \end{aligned}$$

$$= \sigma_{E(XY)} + \frac{3}{4}\sigma_{A(XY)}$$

Similarly, if the data has a hierarchical structure with full-sibs and half-sibs, then the estimated sire, dam and residual covariance components can be expressed in terms of the additive genetic, maternal/common environmental, $\sigma_{M(XY)}$, and environmental covariances in a manner comparable to the analysis of one trait:

$$\sigma_{s(XY)} = \frac{1}{4}\sigma_{A(XY)}$$

$$\sigma_{d(XY)} = \frac{1}{4}\sigma_{A(XY)} + \sigma_{M(XY)}$$

$$\sigma_{e(XY)} = \sigma_{P(XY)} - \frac{1}{2}\sigma_{A(XY)} - \sigma_{M(XY)} \quad \text{or} \quad \begin{aligned} \sigma_{A(XY)} &= 4\sigma_{s(XY)} \\ \sigma_{M(XY)} &= \sigma_{d(XY)} - \sigma_{s(XY)} \\ \sigma_{E(XY)} &= \sigma_{e(XY)} - 2\sigma_{s(XY)} \end{aligned}$$

$$= \sigma_{E(XY)} + \frac{1}{2}\sigma_{A(XY)}$$

Genetic variances and covariance are combined to estimate the genetic correlation

$$r_{A(XY)} = \frac{\sigma_{A(XY)}}{\sqrt{\sigma^2_{A(X)}\sigma^2_{A(Y)}}}$$

and the variance of the genetic correlation is approximately

$$\text{var}(r_{A(XY)}) = \frac{(1-r^2_{A(XY)})^2}{2}\left[\frac{\text{s.e.}(h^2_X)\text{s.e.}(h^2_Y)}{h^2_X h^2_Y}\right]$$

Assuming that the phenotype is the sum of additive genetic and environmental effects, then the phenotypic covariance is the sum of the additive genetic and environmental covariances:

$$\sigma_{P(XY)} = \sigma_{A(XY)} + \sigma_{E(XY)}$$

$$r_P \sigma_{P(X)}\sigma_{P(Y)} = r_A \sigma_{A(X)}\sigma_{A(Y)} + r_E \sigma_{E(X)}\sigma_{E(Y)}$$

$$= r_A h_X h_Y \sigma_{P(X)}\sigma_{P(Y)} + r_E \sqrt{(1-h^2_X)(1-h^2_Y)}\sigma_{P(X)}\sigma_{P(Y)}$$

Dividing by the phenotypic standard deviations for X and Y,

$$r_P = r_A h_X h_Y + r_E \sqrt{(1-h^2_X)(1-h^2_Y)}$$

which combines the phenotypic, genetic and environmental correlations with the heritabilities of the two traits.

Examples of heritabilities, genetic, environmental and phenotypic correlations between growth rate and ultrasonic backfat depth in pigs performance tested on *ad-libitum* or restricted feeding regimes, are shown in Table 3.3.

Table 3.3. Effect of feeding regime on genetic and phenotypic parameters for performance test traits in pigs

Feeding regime	h^2_{ADG}	h^2_{BFAT}	r_A	r_E	r_P
Ad-libitum	0.31	0.50	0.38	0.20	0.27
Restricted	0.17	0.29	-0.10	0.28	0.19

This illustrates that specific genetic and phenotypic parameters are required for a given performance test (Cameron and Curran, 1994; Cameron *et al.*, 1994).

Repeated Measurements

If several measurements are made on each animal, the measurements will not all be the same, due to measurement error, if the measurements were made over a short time period, and also to general environmental variation, if the measurements were taken over a long time interval. Differences between repeated measurements may be due to environmental variation specific to the animal, such as differences between the left- and right-hand sides of a carcass, or if the animal had a lower growth rate, due to a lower food intake compared to another period of time when food intake was normal. The environmental effect can be divided into the general environmental effect, E, which will affect all animals, and a specific environmental effect, E_S, specific to each animal.

The model relating phenotype, genotype and environment is

$$P = G + E_S + E$$

The repeatability of measurements is

$$r_e = \frac{\sigma^2_G + \sigma^2_{E_S}}{\sigma^2_P}$$

such that the repeatability is an upper limit to the heritability.

Example

The data set with eight groups, each with three observations per group, from Chapter 1, is used to illustrate calculation of the repeatability, assuming that the observations were repeat records on each of eight animals. Values in the ANOVA table are the same as before, including the expectation of the mean squares. For the example only, interpretation of the between-animal, σ^2_B, and within-animal variance, σ^2_W, components is $\sigma^2_G + \sigma^2_{E_S}$ and σ^2_E, respectively. The repeatability is 0.19, given the values of 1.25 and 5.33 for σ^2_B and σ^2_W, respectively.

Half-sib Correlation

The half-sib correlation coefficient is estimated in the same manner as the repeatability, in that the estimated between-sire variance component is divided by the sum of the estimated between-sire and within-sire variance components. The repeatability and the half-sib correlation are effectively the correlations between observations in the same sire-family.

The covariance between observations on half-sibs is a quarter of the genetic variance and the variance of observations on half-sibs is the phenotypic variance: then the half-sib correlation coefficient can be written as

$$\frac{\sigma_{XY}}{\sqrt{\sigma_X^2 \sigma_Y^2}} = \frac{\frac{1}{4}\sigma_A^2}{\sqrt{\sigma_P^2 \sigma_P^2}} = \frac{\frac{1}{4}\sigma_A^2}{\sigma_P^2} = \frac{\frac{1}{4}\sigma_A^2}{\frac{1}{4}\sigma_A^2 + \left(\frac{3}{4}\sigma_A^2 + \sigma_E^2\right)} = \frac{\sigma_s^2}{\sigma_s^2 + \sigma_e^2} = \frac{\sigma_B^2}{\sigma_B^2 + \sigma_W^2}$$

The covariance between observations on two half-sibs, HS_1 and HS_2, with respective dams d1 and d2, is

$$\text{cov}(HS_1, HS_2) = \text{cov}\left[\frac{1}{2}(G_s + G_{d1}) + E_1, \frac{1}{2}(G_s + G_{d2}) + E_2\right]$$

$$= \frac{1}{4}\text{cov}(G_s, G_s) = \frac{1}{4}\text{var}(G)$$

$$= \frac{1}{4}\sigma_A^2$$

where G_s and G_d are the genotypes of the sire and dam, while E_1 and E_2 are the environmental effects on each half-sib. No covariance is assumed between the sire and dam or between genotype and environment. For completeness, the covariance between two full-sibs is

$$\text{cov}(FS_1, FS_2) = \text{cov}\left[\frac{1}{2}(G_s + G_d) + E_1, \frac{1}{2}(G_s + G_d) + E_2\right]$$

$$= \frac{1}{4}\text{cov}(G_s, G_s) + \frac{1}{4}\text{cov}(G_d, G_d) = \frac{1}{4}\text{var}(G_s) + \frac{1}{4}\text{var}(G_d)$$

$$= \frac{1}{2}\sigma_A^2 + \sigma_M^2$$

Advantage of repeated measurements

From Chapter 1, the variance of a group mean with n observations per group is $\sigma_B^2 + \sigma_W^2/n$, such that the variance of the mean value of n repeated measurements per animal will be

$$\sigma_G^2 + \sigma_{E_S}^2 + \sigma_E^2/n$$

or

$$r_e\sigma_P^2 + (1 - r_e)\sigma_P^2/n = [r_e + (1 - r_e)/n]\sigma_P^2 \qquad \text{or} \qquad \left[\frac{1 + (n-1)r_e}{n}\right]\sigma_P^2$$

The phenotypic variance of the mean value of n repeated measurements per animal is less than the phenotypic variance of the trait. With repeat measurements, the influence of the specific environmental effect on the mean measurement reduces as the number of observations increases. Therefore, the heritability of the mean measurement is greater than the heritability when one measurement is taken per animal, as the genetic variance accounts for relatively more of the phenotypic variance of the mean measurement.

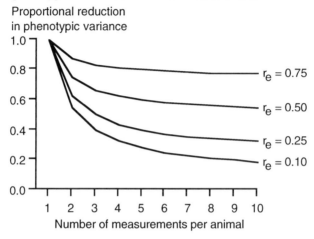

Fig. 3.5. Proportional reduction in the phenotypic variance for the mean of n measurements relative to the phenotypic variance of one measurement

The proportional reduction in the phenotypic variance for the mean of repeated measurements is illustrated in Fig. 3.5. If the repeatability is high, then there is little advantage in increasing the number of observations beyond two, as the impact of the specific environmental effect on one measurement is small. The importance of repeat measurements is discussed in Chapter 4, as the rate of genetic improvement can be increased through the use of repeat measurements.

Chapter four

Identification of Animals of High Genetic Merit

Selection of animals of higher genetic merit than average, to be parents of the next generation, is the basis of genetic improvement programmes, as it is expected that the offspring of selected parents will have higher genetic merit than if the parents had been chosen at random. The mean phenotype of animals in the parental generation, denoted by subscript P, can be written as

$$\bar{P}_P = \bar{G}_P + \bar{E}_P$$

where $\bar{G}$ and $\bar{E}$ are the mean genotypic and environmental effects, and similarly for the mean offspring phenotype, $\bar{P}_O$. The amount of genetic improvement, $\bar{G}_O - \bar{G}_P$, is the difference between the mean offspring phenotype, $\bar{P}_O$, and the mean phenotype of all animals in the parental generation, $\bar{P}_P$, since

$$\bar{P}_O - \bar{P}_P = (\bar{G}_O + \bar{E}_O) - (\bar{G}_P + \bar{E}_P) = \bar{G}_O - \bar{G}_P$$

assuming the same mean environmental contribution to the parental and offspring phenotypes (Fig. 4.1).

Given that the mean phenotype of the parental generation, $\bar{P}_P$, and also the mean phenotype of the animals selected to be parents, $\bar{P}_S$, are known, and if a linear relationship between the phenotypic means

$$(\bar{P}_O - \bar{P}_P) = b(\bar{P}_S - \bar{P}_P)$$

is assumed, then the mean offspring phenotype can be predicted. The assumption of a linear relationship infers a constant rate of change in $(\bar{P}_O - \bar{P}_P)$ as the value of $(\bar{P}_S - \bar{P}_P)$ increases.

The predicted genetic improvement is

$$(\bar{G}_O - \bar{G}_P) = (\bar{P}_O - \bar{P}_P) = b(\bar{P}_S - \bar{P}_P)$$

Prediction of genetic improvement requires knowledge of the regression coefficient relating genotypes to the measured phenotypes, and of the mean phenotypic difference between the animals selected to be parents and all animals in the parental generation. The value of the regression coefficient depends on the relationship between the offspring of selected parents and the individuals whose phenotypic measurements are used to determine which animals in the parental generation are selected. The following section examines how information on an individual can be used to predict its genetic merit.

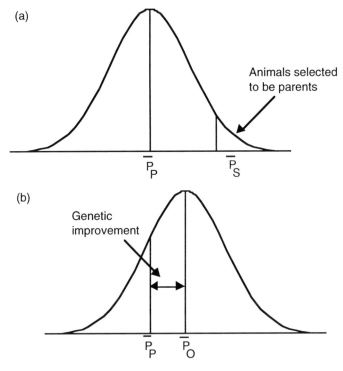

Fig. 4.1. Distribution of phenotypes (a) in the parental generation and (b) of the offspring

Information on the Animal

Single measurement per animal

The true breeding value of an animal is analogous to a population mean, as in both cases the value of the parameter is unknown. To extend the analogy, just as the population mean can be estimated from the mean of a sample of observations from the population, then the true breeding value of an animal can be predicted from a limited number of observations on the individual.

The predicted breeding value, or predicted additive genetic merit, $\hat{A}$, of an animal can be estimated by regression of the animal's breeding value, A, on its phenotype, P (Fig. 4.2).

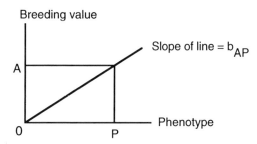

Fig. 4.2. Regression of predicted breeding value on phenotype

With one measurement on the animal, the regression coefficient of additive genetic merit on phenotype is the heritability, since

$$b_{AP} = \frac{\sigma_{AP}}{\sigma_P^2} = \frac{\sigma_A^2}{\sigma_P^2} = h^2$$

The covariance between additive genetic merit and phenotype is equal to the additive genetic variance, assuming that the covariances of additive genetic effect, A, with non-additive genetic effects, nonA, and with environment, E, are zero

$$\text{cov}(A, P) = \text{cov}(A, G + E) = \text{cov}(A, A + \text{nonA} + E) = \text{var}(A)$$

The predicted breeding value of the animal is

$$(\hat{A} - \overline{A}_{Pop}) = h^2 (P - \overline{P}_{Pop})$$

where $\overline{P}_{Pop}$ and $\overline{A}_{Pop}$ are the mean phenotype and mean additive genetic merit of the population. The value of $\overline{A}_{Pop}$ can be set to zero, without any loss of generality. The predicted breeding value, $\hat{A}$, of the animal is

$$\hat{A} = h^2 (P - \overline{P}_{Pop})$$

Repeated measurements per animal

When several measurements have been made on the animal, the predicted breeding value can be determined from the regression coefficient for additive genetic merit on the mean of n measurements, $\overline{P}$. Derivation of the regression coefficient requires both the covariance between the genotype and the mean of n phenotypic measurements and the variance of the mean of n measurements.

Firstly, the covariance is as follows:

$$\text{cov}(A, \overline{P}) = \text{cov}\left(A, \frac{1}{n} \sum_{i=1}^{n} P_i\right) = \frac{1}{n} \sum_{i=1}^{n} \text{cov}(A, P_i) = \text{cov}(A, P)$$

The covariance between the additive genetic merit and the mean of n measurements is equivalent to the covariance between the additive genetic merit and one measurement, which is the additive genetic variance, σ_A^2. The variance of the mean of n measurements, discussed in Chapter 1, is

$$\left[\frac{1+(n-1)r_e}{n}\right]\sigma_P^2$$

where r_e is the repeatability of the measurement. The regression coefficient for genotype on the mean of n measurements is

$$b_{A\bar{P}} = \left[\frac{n}{1+(n-1)r_e}\right]\frac{\sigma_A^2}{\sigma_P^2} = \left[\frac{nh^2}{1+(n-1)r_e}\right]$$

The predicted genetic merit of an animal with n measurements is

$$\hat{A} = \left[\frac{nh^2}{1+(n-1)r_e}\right]\left(\bar{P} - \bar{P}_{Pop}\right)$$

The manner in which the regression coefficient weights the mean of n measurements is easier to appreciate if it is assumed that the repeatability is equal to the heritability and that the phenotypic mean of the population is zero. Then, the predicted genetic merit is

$$\hat{A} = \left[\frac{nh^2}{1+(n-1)h^2}\right]\bar{P} = \left[\frac{n}{n+\lambda}\right]\bar{P}$$

where $\lambda = \frac{1-h^2}{h^2} = \frac{\sigma_E^2}{\sigma_A^2}$.

The regression coefficient increases as the number of measurements and the heritability increase, such that more weight is given to the phenotypic mean measurement for prediction of genetic merit.

Before illustrating the advantage of repeated measurements in genetic improvement programmes with an example, the selection differential and response to selection need to be defined.

Selection Differential

If animals in the parental generation are ranked according to predicted genetic merit, and a given proportion of those with highest genetic merit are selected to be parents, then the response to selection can be predicted, as the difference in predicted genetic merit of the offspring and of the parental generation. As the proportion of animals selected to be parents decreases, the mean predicted genetic merit of selected animals increases and the genetic improvement also increases.

The predicted response to selection depends on both the proportion of animals selected and the regression coefficient of additive genetic merit on phenotype. Assuming that phenotypic measurements are normally distributed and that a proportion, p, of animals with extreme phenotypes are selected, then it is expected that the minimum phenotype of the selected animals will exceed the mean phenotype of the parental generation by x phenotypic standard deviations, $x\sigma_P$.

Identification of animals of high genetic merit

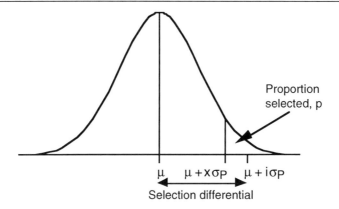

Fig. 4.3. Selection differential

The expected difference between the mean phenotype of the selected animals, $\mu + i\sigma_P$, and the mean phenotype of the parental generation, μ, is the selection differential, $i\sigma_P$, measured in trait units (Fig. 4.3). The parameters i and x are termed the standardised selection differential and the truncation point. For a given value of p, the values of i and x can be obtained from normal distribution tables (see Appendix).

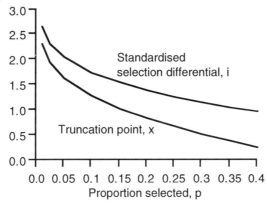

Fig. 4.4. Standardised selection differential and truncation point

In Chapter 1, some properties of the normal distribution were discussed; in particular, that a defined proportion of observations are expected to have values between the mean and a specific number of standard deviations from the mean. For example, 0.34 of the observations are expected to lie between the mean and one standard deviation above the mean. Conversely, 0.16 of the observations are expected to have values greater than one standard deviation above the mean, with an average value greater than the mean by 1.53 standard deviations (Fig. 4.4).

Expected values for the selection differential, in standard deviation units, for various proportions of selected animals, are given in the Appendix.

If sires and dams have different selection differentials, i_S and i_D, respectively, then the selection differential of parents is

$$i = \frac{i_S + i_D}{2}$$

Response to Selection

The response to selection, R, is the difference between the mean phenotypes of the progeny and parental generations, which can be predicted given the selection differential, SD, and the regression coefficient relating genotype to phenotype.

When each animal has one measurement, the regression coefficient is the heritability and the response to selection is

$$R = h^2 SD = ih^2 \sigma_P$$

When there are n measurements on each animal, the selection differential is measured in terms of $\sigma_{\bar{P}}$, rather than in phenotypic standard deviations, since the animals are selected on the basis of the mean of n measurements. The response to selection is

$$R_n = b_{A\bar{P}} SD = \left[\frac{nh^2}{1+(n-1)r_e}\right] i\sigma_{\bar{P}} = ih^2 \sigma_P \sqrt{\frac{n}{1+(n-1)r_e}}$$

as the variance of n measurements is $\left[\dfrac{1+(n-1)r_e}{n}\right]\sigma_P^2$.

The advantage of repeated measurements is illustrated in terms of the response, R_n, relative to the response with one measurement, R, per animal (see Fig. 4.5). The relative response depends on the repeatability and the number of measurements, but not on the heritability of the trait.

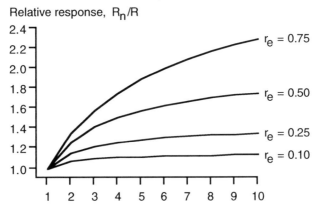

Fig. 4.5. Increase in response with repeated measurements per animal

There is a substantial benefit of having repeated measurements when the repeatability is low, but not when the repeatability is high. With six measurements per animal and a repeatability of 0.1, the response is d͏͏͏c when the repeatability is 0.75, the proportional increase in the respoɩ͏ɪ͏ᴄ ᴜ͏ ͏ᴜ᠁ᴊ 0.1 (Fig. 4.5). In many situations, it will not be possible to take multiple measurements, but even with only two measurements per animal, there are substantial increases in the response to selection, irrespective of the repeatability.

Before discussing the relative advantages and disadvantages of alternative selection procedures, in a later section of this chapter, several parameters relating to the predicted breeding value will be described.

Variance of Predicted Breeding Value

The variance of a predicted breeding value is a measure of the precision with which the breeding value is estimated. Calculation of the variance of a predicted breeding value uses several properties of the variance, described in Chapter 1:

$$\text{var}(\hat{A}) = \text{var}\left[b_{A\bar{P}}(\bar{P} - \bar{P}_{Pop})\right]$$

$$= \text{var}(b_{A\bar{P}}\bar{P}) \quad \text{regarding } \bar{P}_{Pop} \text{ as a constant and } \text{var}(X+\alpha) = \text{var}(X)$$

$$= b_{A\bar{P}}^2 \text{var}(\bar{P}) \quad \text{regarding } b_{A\bar{P}} \text{ as a constant and } \text{var}(\alpha X) = \alpha^2 \text{var}(X)$$

$$= b_{A\bar{P}} \text{cov}(A, \bar{P}) \quad \text{since } b_{A\bar{P}} = \text{cov}(A, \bar{P})/\text{var}(\bar{P})$$

$$= \frac{nh^4}{1+(n-1)r_e} \sigma_P^2 \quad \text{as } \text{cov}(A, \bar{P}) = \text{cov}(A, P) = \sigma_A^2 = h^2 \sigma_P^2$$

Again, it is useful to assume that the repeatability is equal to the heritability, to appreciate how changes in the number of measurements and in the heritability affect the variance of a predicted breeding value.

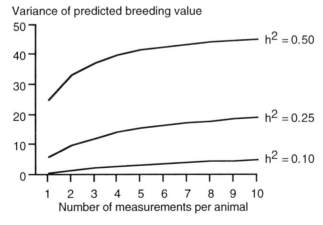

Fig. 4.6. Change in variance of predicted breeding value with repeated number of measurements per animal, assuming that $h^2 = r_e$

In Fig. 4.6, a phenotypic variance of 100 has been assumed. As the number of measurements per animal increases, the variance of the predicted breeding value increases and tends towards the additive genetic variance. The variance of a predicted breeding value behaves in the opposite manner to the variance of an estimated group mean, which decreases towards the between-group variance as the precision of the estimate increases.

Assuming that the repeatability equals the heritability, the variance of the predicted breeding value can be expressed as

$$\text{var}(\hat{A}) = \frac{n}{n+\lambda} \sigma_A^2$$

where $\lambda = \dfrac{1-h^2}{h^2} = \dfrac{\sigma_E^2}{\sigma_A^2}$

Accuracy of Predicted Breeding Value

The accuracy of a predicted breeding value is the correlation between the true breeding value and the predicted breeding value, and, with the variance of the predicted breeding value, is a another measure of the precision with which the breeding value has been estimated. Although the true breeding value is not known, the additive genetic variance is known, and the covariance between the true and predicted breeding values can be quantified.

The covariance between the true and predicted breeding values is

$$\text{cov}(A, \hat{A}) = \text{cov}(A, b_{A\bar{P}}\bar{P}) = b_{A\bar{P}} \text{cov}(A, \bar{P})$$

and since $b_{A\bar{P}} \text{cov}(A, \bar{P})$ is the variance of the predicted breeding value, then the correlation between the true and predicted breeding values is

$$r_{A\hat{A}} = \frac{\text{cov}(A, \hat{A})}{\sqrt{\text{var}(A)\text{var}(\hat{A})}} = \sqrt{\frac{\text{var}(\hat{A})}{\text{var}(A)}} = h\sqrt{\frac{n}{1+(n-1)r_e}}$$

The square of the correlation between the true and predicted breeding values is just the regression coefficient with which the breeding value is predicted. If the repeatability is equal to the heritability, then as the number of repeated measurements increases, the accuracy of the predicted breeding value tends to unity.

It should be noted that the regression of the true breeding value on the predicted breeding value is

$$\frac{\text{cov}(A, \hat{A})}{\text{var}(\hat{A})} = \frac{\text{var}(\hat{A})}{\text{var}(\hat{A})} = 1$$

which is appropriate, since the predicted breeding value predicts the true breeding value.

Rearranging the formula for the square of the accuracy and assuming that the repeatability is equal to the heritability, then

$$r^2_{A\hat{A}} = \frac{n}{n+\lambda}$$

where $\lambda = \dfrac{1-h^2}{h^2} = \dfrac{\sigma^2_E}{\sigma^2_A}$

When the repeatability and heritability are equal, the accuracy of a predicted breeding value can be quickly determined. For example, if the heritability and repeatability equal 0.25, then the value of λ is 3, and the accuracy of the predicted breeding value for an animal with 1, 2, 4 or 6 repeated measurements is 0.50, 0.63, 0.76 and 0.82, respectively.

The accuracy of the predicted breeding value can also be used to predict the response to selection, given the selection differential. The response to selection was previously defined as the product of the regression coefficient of additive genetic merit on phenotype with the selection differential:

$$R_n = ih^2\sigma_P \sqrt{\frac{n}{1+(n-1)r_e}}$$

The regression coefficient is the square of the accuracy of the predicted breeding value, so the response can be expressed in terms of the accuracy

$$R_n = ih\sigma_P\, r_{A\hat{A}} \qquad \text{from the definition of } r_{A\hat{A}}$$
$$= i r_{A\hat{A}} \sigma_A$$

Given the accuracy of the predicted breeding value and the standardised selection differential, the response to selection can be predicted.

Prediction Error Variance

The prediction error variance (PEV) is a measure of the variation about the predicted breeding value, or is the variation in the mean phenotypic measurement of the individual which is not accounted for by the regression of additive genetic merit on phenotype. The prediction error variance is

$$\mathrm{var}(A \text{ about } \hat{A}) = \left(1 - r^2_{A\hat{A}}\right)\mathrm{var}(A)$$
$$= \mathrm{var}(A) - r^2_{A\hat{A}}\,\mathrm{var}(A)$$
$$= \mathrm{var}(A) - \mathrm{var}(\hat{A}) \qquad \text{since } r^2_{A\hat{A}} = \frac{\mathrm{var}(\hat{A})}{\mathrm{var}(A)}$$

Derivation of the prediction error variance provides the result that the additive genetic variance is the sum of the variance of the predicted breeding value and the prediction error variance. As the number of measurements per animal increases, the prediction error variance decreases, while the variance of the predicted breeding value increases.

Again, assuming the repeatability and heritability are equal, for purposes of illustration, the variance of predicted breeding value complements the prediction error variance, as shown:

Genetic variance	Variance of predicted breeding value	Prediction error variance
σ_A^2	$\dfrac{n}{n+\lambda}\sigma_A^2$	$\dfrac{\lambda}{n+\lambda}\sigma_A^2$

In Fig. 4.7, the variance of the predicted breeding value and the prediction error variance were calculated for a heritability of 0.1 and a phenotypic variance of 100, assuming that the heritability and repeatability were equal. The complementary nature of the two parameters is obvious, with the prediction error variance decreasing as the number of measurements per animal increases, and the variance of the predicted breeding value tending to the additive genetic variance. Although animals are unlikely to ever have 100 measurements, up to 100 measurements were assumed to illustrate the complementary nature of the variance of the predicted breeding value and the prediction error variance.

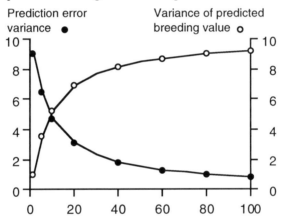

Fig. 4.7. The relationship between the prediction error variance and the variance of the predicted breeding value

The prediction error variance is also used to derive the confidence interval for the predicted breeding value. The limits of the $(1-2\alpha)$ confidence interval are $\hat{A} \pm x\sqrt{\text{PEV}}$, where x is the value from the normal distribution, measured in standard deviation units, such that a proportion α of observations are greater than x, and PEV is the prediction error variance. Values of x corresponding to proportions, p, are given in the Appendix.

Example
The average weaning weight of Florence's first litter of piglets was 51.5 kg, while Emily had eight litters that averaged 46.0 kg. Predict the breeding values, the prediction error variance, the accuracy of prediction and the 0.95 confidence intervals for Florence and Emily, given that litter weaning weight has a heritability of 0.3, a repeatability of 0.4 and a phenotypic variance of $16\,\text{kg}^2$,

and that the population mean is 41 kg. Litter weaning weight is considered to be a trait of the dam.

		Florence	Emily
Breeding value	$\hat{A} = b_{A\bar{P}}(\bar{P} - \bar{P}_{Pop})$	3.15	3.16
Accuracy	$r_{A\hat{A}} = \sqrt{b_{A\bar{P}}}$	0.55	0.79
Prediction error variance	$PEV = (1 - r_{A\hat{A}}^2)\sigma_A^2$	3.36	1.77
Confidence interval	$\hat{A} \pm 1.96\sqrt{PEV}$	(−0.44, 6.74)	(0.55, 5.76)

The higher mean litter weight of Florence was compensated by the lower number of litters than Emily, as the breeding values of the two sows were the same. However, the larger PEV of Florence meant that her breeding value was not significantly different from zero. The accuracy of Emily's predicted breeding value is higher than for Florence, so which animal should be selected?

The higher accuracy of Emily's predicted breeding value means a smaller confidence interval than for Florence. The probabilities that Emily's and Florence's true breeding values are greater than 5.76 are 0.025 and 0.075, as 5.76 is 1.43 prediction error standard deviations greater than Florence's predicted breeding value of 3.15. To the entrepreneur, the probability of 0.075, that Florence's true breeding value is greater than the upper confidence interval of Emily's true breeding value, may be sufficient to offset the probability of 0.042, that Florence's true breeding value is actually less than zero. However, the probability that Florence's true breeding value is less than zero is nearly five times greater than the probability of Emily's true breeding value being less than zero, which may convince the conservative person to select Emily.

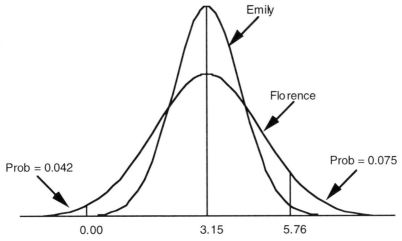

Fig. 4.8. Distributions of predicted breeding values for Florence and Emily

Chapter five

Information from Relatives

Information from Sibs

Measurements on an individual's relatives can contribute information towards the individual's predicted breeding value, as measurements on sibs are comparable with repeated measurements on the individual, once account has been taken of the genetic relationship between the individual and its sibs. Prediction of the individual's breeding value from measurements on its n sibs requires the regression coefficient of the individual's breeding value on the mean value of its sibs' measurements. The variance of the mean of n measurements is

$$\left[\frac{1+(n-1)t}{n}\right]\sigma_P^2 = \left[t + \frac{1-t}{n}\right]\sigma_P^2$$

where t is the repeatability, or the correlation between sibs. If the individual's relatives are half-sibs, the progeny of a sire mated to a random group of dams and each dam has only one offspring, then $t = \frac{1}{4}h^2$, but if the individual's relatives are all full-sibs, then $t = \frac{1}{2}h^2 + c^2$, where h^2 and c^2 are the heritability and the maternal effect, which includes the common environmental effect, respectively. Calculation of the regression coefficient requires the covariance between the individual's breeding value and the mean of its sibs' measurements. However, the covariance depends on which relatives are measured and whether or not the individual is measured. Therefore, each case must be considered separately.

Selection on sib information

With selection on sib information, the breeding value of the individual is predicted from the measurements of its sibs, but the individual is not measured. Selection decisions are made on the sib of the animals with measurements, rather than on the animals with measurements. Measurements on sibs can be used to provide information on reproductive traits, carcass composition or meat quality, which could not be obtained on potential breeding animals.

In the section on repeated measurements on the individual, in Chapter 4, the covariance between genotype and mean of n measurements is equivalent to the covariance between genotype and one measurement. Therefore, the covariance between an individual's breeding value and its sibs' mean measurement is the covariance between an individual's breeding value and one sib's measurement.

The covariance is $r\sigma_A^2$, where r is the genetic relationship between the individual and its relative, which is a half for full-sibs and a quarter for half-sibs. The genetic relationship between two animals is the probability that the genotypes of the two animals, for a gene taken at random, are identical by descent. The value of the genetic relationship can be simply determined from the pedigree, by counting the number of steps in the pedigree that connect the individual to the relative via the sire and via the dam. The genetic relationship is then one half to the power of the number of sire steps plus a half to the power of the number of dam steps.

For example, with an individual and its half-sib, two steps connect the individual to the half-sib via the sire, but no steps connect via the dam, as they do not have a dam in common, so the genetic relationship is $\left(\frac{1}{2}\right)^2 = \frac{1}{4}$:

For the individual and a full-sib, two steps connect the individual to the full-sib via the sire and via the dam, so the genetic relationship is $\left(\frac{1}{2}\right)^2 + \left(\frac{1}{2}\right)^2 = \frac{1}{2}$.

Therefore, the regression coefficient of the individual's breeding value on the mean measurement of its n sibs is

$$b_{A\overline{S}} = \frac{r\sigma_A^2}{\left[\frac{1+(n-1)t}{n}\right]\sigma_P^2} = \frac{nrh^2}{1+(n-1)t}$$

which has similar form to the regression coefficient for repeated measurements on the individual of

$$b_{A\overline{P}} = \frac{nh^2}{1+(n-1)r_e}$$

with rh^2 replacing h^2 to account for the genetic relationship between the individual and its sibs and the correlation, t, between the sibs' measurements replacing the repeatability, r_e, of the individual's measurements.

With selection on sib information, the predicted breeding value of the individual is

$$\hat{A} = b_{A\overline{S}}\left(\overline{S} - \overline{P}_{Pop}\right)$$

where $\bar{S}$ is the mean measurement of the n sibs and $\bar{P}_{Pop}$ is the mean phenotype of the population.

When measurements are taken on sibs, then the square of the accuracy of the predicted breeding value is obtained from:

$$r^2_{A\hat{A}} = \frac{\text{var}(\hat{A})}{\text{var}(A)} = \frac{b^2_{A\bar{S}}\,\text{var}(\bar{S})}{\text{var}(A)}$$

where $\bar{S}$ represents the mean sib measurement, as discussed in Chapter 4. Therefore, the square of the accuracy is:

$$r^2_{A\hat{A}} = \frac{nr^2 h^2}{1+(n-1)t}$$

When measurements are taken only on half-sibs, then $r = \frac{1}{4}$ and $t = \frac{1}{4}h^2$ and the square of the accuracy of the predicted breeding value simplifies to

$$r^2_{A\hat{A}} = \frac{1}{4}\frac{n}{n+\lambda}$$

where $\lambda = \dfrac{4-h^2}{h^2}$

Selection on between-family deviations

With selection on between-family deviations, the predicted breeding value of the family is based on the deviation between the mean measurement of all animals in the family and the population mean. Therefore, all n family members have the same predicted breeding value and the whole family is either selected or not, as there is no differentiation between family members. Although all family members are allocated the same predicted breeding value, derivation of the regression coefficient of breeding value on the mean measurement of the family is based on one family member. The covariance between the individual's breeding value and mean measurement of the family has to account for the individual being included in the mean. The individual's phenotype and breeding value are denoted P_1 and A_1, with the (n−1) phenotypes of the individual's family members denoted P_i, for i equal to 2,3...,n. The covariance between the individual's breeding value and mean measurement of the family, $\bar{F}$, is

$$\text{cov}(A_1, \bar{F}) = \text{cov}\left[A_1, \frac{1}{n}\sum_{i=1}^{n} P_i\right]$$

$$= \frac{1}{n}\text{cov}(A_1, P_1) + \frac{1}{n}\sum_{i=2}^{n}\text{cov}(A_1, P_i)$$

$$= \frac{1}{n}\sigma^2_A + \frac{n-1}{n} r\sigma^2_A$$

since the covariance between the individual's breeding value and phenotype is the additive genetic variance, while the covariance between the individual's breeding value and the phenotype of a family member is the additive genetic variance multiplied by the genetic relationship. Therefore,

$$\text{cov}(A_1, \bar{F}) = \left[\frac{1+(n-1)r}{n}\right]\sigma_A^2$$

The variance of the mean of n measurements is

$$\left[\frac{1+(n-1)t}{n}\right]\sigma_P^2$$

so the regression coefficient of breeding value on mean family measurement is

$$b_{A\bar{F}} = \left[\frac{1+(n-1)r}{1+(n-1)t}\right]h^2$$

With selection on between-family deviations, the predicted breeding value of each family member is

$$\hat{A} = b_{A\bar{F}}(\bar{F} - \bar{P}_{Pop})$$

where $\bar{F}$ is the mean measurement of the family and $\bar{P}_{Pop}$ is the mean phenotype of the population.

Selection on within-family deviations

In selection on between-family deviations, all members of a family are selected, as there is no discrimination between family members. Conversely, for selection on within-family deviations, animals are assessed relative to the family mean, such that information on between-family differences is not used in the selection decision. Animals are selected solely on the basis of their deviation from their family mean, with no account taken of the family group, such that the number of animals selected from each family will be variable (Hill et al., 1996). Selection on within-family deviations would be used when there are large environmental effects specific to each family. Direct comparison of phenotypes of animals from different families would include comparison of the between-family environmental effects, such that differences in phenotypes would not directly reflect differences in additive genetic merit. For example, the phenotypic difference between an animal in family 1 and an animal in family 2, assuming that the general environment is similar to both animals, is

$$P_1 - P_2 = (A_1 + M_1 + E_1) - (A_2 + M_2 + E_2) = (A_1 - A_2) + (M_1 - M_2)$$

where M is the maternal and common environmental effect

For selection on within-family deviations, an animal's predicted breeding value is

$$\hat{A} = b_{AP}(P - \bar{F})$$

where P is the animal's phenotype and $\bar{F}$ is the mean family measurement, which includes the animal's measurement. Derivation of the regression coefficient for the animal's breeding value on the deviation between the animal's measurement and the family mean requires the covariance between the animal's breeding value and the difference between the animal's measurement and the family mean,

$\mathrm{cov}(A, P - \overline{F})$, and the variance of the difference between the animal's measurement and the family mean, $\mathrm{var}(P - \overline{F})$.

As in derivation of the regression coefficient for selection on between-family deviations, it is assumed that the individual's phenotype and breeding value are P_1 and A_1, with the (n−1) phenotypes of the individual's family members equal to P_i, for i = 2,3...,n.. The covariance is

$$\mathrm{cov}(A_1, P_1 - \overline{F}) = \mathrm{cov}(A_1, P_1) - \mathrm{cov}(A_1, \overline{F})$$

$$= \sigma_A^2 - \left[\frac{1}{n}\sigma_A^2 + \frac{n-1}{n} r\sigma_A^2\right]$$

$$= \left(1 - \frac{1}{n}\right)(1-r)\sigma_A^2$$

The variance of the difference between the animal's measurement and the family mean is

$$\mathrm{var}(P_1 - \overline{F}) = \mathrm{var}(P_1) + \mathrm{var}(\overline{F}) - 2\,\mathrm{cov}(P_1, \overline{F})$$

which uses property (5) of the variance, that

$$\mathrm{var}(cX + dY) = c^2\,\mathrm{var}(X) + d^2\,\mathrm{var}(Y) + 2cd\,\mathrm{cov}(X, Y)$$

The variance of the animal's measurement, $\mathrm{var}(P_1)$, and of the mean family measurement, $\mathrm{var}(\overline{F})$, have already been determined, so only the covariance between the measurements of the animal and the family mean remains:

$$\mathrm{cov}(P_1, \overline{F}) = \mathrm{cov}\left(P_1, \frac{1}{n}\sum_{i=1}^{n} P_i\right)$$

$$= \frac{1}{n}\mathrm{cov}(P_1, P_1) + \frac{1}{n}\sum_{i=2}^{n}\mathrm{cov}(P_1, P_i)$$

$$= \frac{1}{n}\sigma_P^2 + \frac{n-1}{n}t\sigma_P^2$$

as the covariance between the measurement of the animal and a family member is phenotypic variance multiplied by the intra-family correlation, t.

The variance of the difference between the animal's measurement and the family mean is

$$\mathrm{var}(P_1 - \overline{F}) = \mathrm{var}(P_1) + \mathrm{var}(\overline{F}) - 2\,\mathrm{cov}(P_1, \overline{F})$$

$$= \sigma_P^2 + \left[t + \frac{1-t}{n}\right]\sigma_P^2 - 2\left[\frac{1}{n}\sigma_P^2 + \frac{n-1}{n}t\sigma_P^2\right]$$

$$= \left(1 - \frac{1}{n}\right)(1-t)\sigma_P^2$$

The regression coefficient for the animal's breeding value on the difference between the animal's measurement and the mean family measurement is

$$b_{AP} = \left(\frac{1-r}{1-t}\right)h^2$$

For selection on within-family deviations, the predicted breeding value of the individual is

$$\hat{A} = b_{AP}(P - \bar{F})$$

where $\bar{F}$ is the mean family measurement of the n family members, including the individual.

Responses to selection on between-family and within-family deviations

Responses to selection on between-family and within-family deviations are equal to the product of the standardised selection differential, i, the appropriate regression coefficient to predict breeding values from measurements, b, and the standard deviation, σ, of the actual trait on which selection is based. Direct comparison of the two selection methods is difficult, as the trait on which selection is practised and the corresponding standard deviations differ. In general, the standardised selection differentials will be quite different for selection on between-family or within-family deviations, as the maximum family size may be lower than the number of families. The selection methods can be compared by expressing the responses as multiples of $ih^2\sigma_P$:

Between-family $\quad \dfrac{r + \dfrac{1-r}{n}}{\sqrt{t + \dfrac{1-t}{n}}} \quad$ which tends to $\dfrac{r}{\sqrt{t}}$ as n tends to ∞

Within-family $\quad (1-r)\sqrt{\dfrac{1-\dfrac{1}{n}}{1-t}} \quad$ which tends to $\dfrac{1-r}{\sqrt{1-t}}$ as n tends to ∞

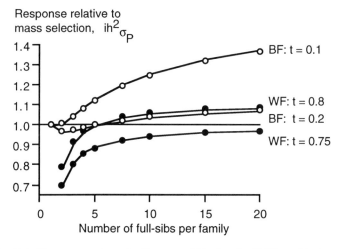

Fig. 5.1. Responses to selection on within-family (WF) and between-family (BF) deviations, relative to the response with mass selection

With large numbers of full-sibs, the response to selection on within-family deviations, relative to mass selection, increases as the full-sib correlation increases, while the response to selection on between-family deviations decreases (Fig. 5.1). Assuming the same standardised selection differential and a low value for the correlation between full-sibs, then the response to selection on between-family deviations will be relatively greater than the response to selection on within-family deviations.

Information from Progeny

Measurements on an individual's relatives are not just restricted to full-sibs or half-sibs, as information from progeny can also be used to predict an individual's breeding value. Progeny testing is standard practice in the dairy industry to obtain milk quantity and quality information, for prediction of breeding values for bulls. Similarly, progeny test information can be used to collect data on reproductive performance or on carcass composition.

Incorporation of progeny test information to predict the animal's breeding value is comparable to the use of repeated measurements on the individual, once the genetic relationship between the individual and its progeny is accounted for. It is straightforward to derive the regression coefficient of the animal's breeding value on the mean of its progeny measurements. The variance of the progeny mean, $\overline{P}$, is just the variance of the mean of n measurements, which is

$$\text{var}(\overline{P}) = \left[\frac{1+(n-1)t}{n}\right]\sigma_P^2$$

where t is the between-progeny correlation. For full-sib progeny, the correlation is $\frac{1}{2}h^2 + c^2$ and for half-sib progeny the correlation is $\frac{1}{4}h^2$. The covariance between the individual and the mean of its progeny measurements is equivalent to the covariance between the individual and one of its progeny, which is $\frac{1}{2}\sigma_A^2$, as discussed when predicting an animal's breeding value from the mean of its n sib measurements. The regression coefficient of predicted breeding value on the mean of n half-sib progeny measurements, $\overline{P}$, is

$$b_{A\overline{P}} = \frac{\frac{1}{2}\sigma_A^2}{\left[\frac{1+(n-1)t}{n}\right]\sigma_P^2} = \frac{rnh^2}{1+(n-1)t}$$

since $r = \frac{1}{2}$, and the animal's predicted breeding value is

$$\hat{A} = b_{A\overline{P}}(\overline{P} - P_{Pop})$$

where P_{Pop} is the mean phenotype of the population.

Rearranging the equation for the regression coefficient gives

$$b_{A\overline{P}} = \frac{2n}{n+\lambda} \qquad \text{where } \lambda = \frac{4-h^2}{h^2}$$

As the number of progeny measurements increases, the regression coefficient tends to 2, since the genetic merit of the individual's progeny reflect only half the genetic merit of the individual.

The regression coefficient of the individual's breeding value on the mean measurement of its n progeny is of the same form as the regression coefficient when measurements are available on the individual's sibs

$$b_{A\bar{S}} = \frac{rnh^2}{1+(n-1)t}$$

and when repeated measurements are made on the individual itself

$$b_{A\bar{P}} = \frac{nh^2}{1+(n-1)r_e}$$

with firstly, rh^2 replacing h^2 to account for the genetic relationship between the individual and its sibs or progeny and, secondly, the intra-class correlation, t, between the sibs' or progeny measurements replacing the repeatability, r_e, of the individual's measurements.

Given progeny information, the square of the accuracy of the individual's predicted breeding value is derived from:

$$r^2_{A\hat{A}} = \frac{\text{var}(\hat{A})}{\text{var}(A)} = \frac{b^2_{A\bar{P}}\,\text{var}(\bar{P})}{\text{var}(A)}$$

The square of the accuracy is:

$$r^2_{A\hat{A}} = \frac{nr^2h^2}{1+(n-1)t}$$

Since $r = \frac{1}{2}$ and for half-sib progeny $t = \frac{1}{4}h^2$, then:

$$r^2_{A\hat{A}} = \frac{n}{n+\lambda} \qquad \text{where } \lambda = \frac{4-h^2}{h^2}$$

Note that the accuracy of predicting the individual's breeding value with half-sib measurements is half the accuracy with measurements on half-sib progeny, since half-sib measurements predict the breeding value of the individual's sire.

Response with selection on progeny information

The equation for the response to selection given progeny information is

$$R = \frac{1}{2}ih^2\sigma_P\sqrt{\frac{n}{1+(n-1)t}}$$

which is similar to the response with repeated measurements on the individual

$$R = ih^2\sigma_P\sqrt{\frac{n}{1+(n-1)r_e}}$$

with the repeatability replaced by the intra-class correlation between progeny and account taken of the genetic relationship between the individual and its progeny.

Information from Parents

An animal's breeding value can also be estimated from the predicted breeding values of its parents, as

$$\hat{A} = \frac{1}{2}(\hat{A}_S + \hat{A}_D)$$

where $\hat{A}_S$ and $\hat{A}_D$ are the sire's and dam's predicted breeding values, respectively. The response to selection using the average parental predicted breeding value is

$$R = b_{A\hat{A}} SD_{\hat{A}}$$

where $SD_{\hat{A}}$ is the selection differential for average parental predicted breeding values. Calculation of the regression coefficient of predicted breeding value on average parental predicted breeding value requires the covariance between breeding value with average parental predicted breeding value, $\text{cov}(A, \hat{A})$ and the variance of average parental predicted breeding values, $\text{var}(\hat{A})$.

The covariance is

$$\text{cov}(A, \hat{A}) = \text{cov}\left[\frac{1}{2}(A_S + A_D), \frac{1}{2}(\hat{A}_S + \hat{A}_D)\right]$$

$$= \frac{1}{4}\left[\text{cov}(A_S, \hat{A}_S) + \text{cov}(A_D, \hat{A}_D)\right]$$

Assuming the sire and dam breeding values are uncorrelated, the the covariance is

$$= \frac{1}{4}\left[r_S^2 + r_D^2\right]\sigma_A^2 \quad \text{since } r_{A\hat{A}}^2 = \frac{\text{var}(\hat{A})}{\text{var}(A)}, \text{ from Chapter 4}$$

Similarly, the variance of the average of the parental predicted breeding values is

$$\text{var}(\hat{A}) = \text{var}\left[\frac{1}{2}(\hat{A}_S + \hat{A}_D)\right]$$

$$= \frac{1}{4}\left[r_S^2 + r_D^2\right]\sigma_A^2$$

The regression coefficient of predicted breeding value on average parental predicted breeding value is unity, so the response to selection is just the selection differential for average parental predicted breeding values

$$b_{A\hat{A}} SD_{\hat{A}} = i\sigma_{\hat{A}} = \frac{i}{2}\sqrt{(r_S^2 + r_D^2)}\sigma_A = \frac{i}{2}\sqrt{(r_S^2 + r_D^2)}h\sigma_P$$

As the accuracy of the parental predicted breeding values increases, then the response to selection on average parental predicted breeding value increases.

The accuracy of the animal's breeding value estimated from the predicted breeding values of its parents is $\frac{1}{2}\sqrt{(r_S^2 + r_D^2)}$. If the accuracy of each parent's predicted breeding value was unity, then the accuracy of the animal's predicted breeding value would be $\sqrt{\frac{1}{2}}$, which is the same as if the breeding value had been predicted from a very large number of full-sibs. The upper limit of the accuracy is constrained to $\sqrt{\frac{1}{2}}$, due to the effect of Mendelian sampling, as the animal's two alleles at a particular locus are not equal to both alleles of either parent.

Example

An animal's sire has 50 progeny test records and the dam has two repeated measurements. If the heritability of the trait is 0.25, then what is the accuracy of the animal's predicted breeding value, assuming that the heritability and repeatability are equal?

The accuracies of the sire's and dam's predicted breeding values are

$$\sqrt{\frac{n}{n+\frac{4-h^2}{h^2}}} = 0.88 \quad \text{and} \quad \sqrt{\frac{n}{n+\frac{1-h^2}{h^2}}} = 0.63$$

The accuracy of the animal's predicted breeding value is $\frac{1}{2}\sqrt{\left(r_S^2 + r_D^2\right)} = 0.54$.

Predicting breeding values with measurements on the animal, sibs and progeny

There are several similarities between predicting an animal's breeding value from repeated measurements on itself compared to measurements on its sibs and progeny. For example, the regression coefficients for breeding value prediction have a similar structure, with the correlation between sibs or progeny replaced by the repeatability and inclusion of the genetic relationship between the animal and its sibs or progeny. The square of the accuracy of the predicted breeding value is the regression coefficient multiplied by the genetic relationship between the animal and its sibs or progeny. If the repeatability is equal to the heritability and there is no maternal or common environmental effect, then as the number of measurements on the individual or the number of half-sib progeny increase, the accuracy tends to unity, but with measurements on full-sibs, the accuracy tends to $\sqrt{\frac{1}{2}}$ or 0.71 (see Table 5.1). The difference in the upper limits of the accuracies between the mean full-sib and mean progeny measurements is due to Mendelian sampling of alleles. Mendelian sampling can be thought of as the random process of allocating parental alleles to progeny. For example, if the sire's and dam's genotypes at a particular locus are AB and CD, then the four possible progeny genotypes are AC, AD, BC and BD. If one offspring has genotype AD, then some of its full-sibs may have a completely different genotype, BC, or only have half the alleles in common, AC or BD. Measurements on full-sibs are not necessarily the most reliable indication of the animal's genotype. However, all of the sire's progeny have one of his alleles, such that the mean progeny measurement is a more reliable predictor of the sire's genotype than the mean of the sire's full-sibs.

Table 5.1. Formula for the regression coefficient of predicted breeding value on mean phenotypic measurement and the accuracy of selection

	Repeated measurements	Measurements on full-sibs	Measurements on half-sib progeny
Regression coefficient, b	$\dfrac{nh^2}{1+(n-1)r_e}$	$\dfrac{\frac{1}{2}nh^2}{1+(n-1)t}$	$\dfrac{\frac{1}{2}nh^2}{1+(n-1)\frac{1}{4}h^2}$
(Accuracy of selection)²	b	$\frac{1}{2}b$	$\frac{1}{2}b$
If repeatability = heritability and maternal effect = 0, then (accuracy of selection)²	$\dfrac{n}{n+\dfrac{1-h^2}{h^2}}$	$\dfrac{\frac{1}{2}n}{n+\dfrac{2-h^2}{h^2}}$	$\dfrac{n}{n+\dfrac{4-h^2}{h^2}}$

Responses with Measurements on the Animal, Sibs and Progeny

The response to selection was defined as the regression coefficient of breeding value on measurements, b, multiplied by the selection differential, SD. The regression coefficient is a multiple of the square of the accuracy of the predicted breeding value, so the response can be expressed in terms of the accuracy.

For example, with sib selection, the response is as follows:

$$\text{response} = b_{A\overline{S}}(\overline{S} - P_{Pop})$$

$$= \dfrac{\sigma_{A\overline{S}}}{\sigma_{\overline{S}}^2} i\sigma_{\overline{S}} \qquad \text{since the selection differential} = i\sigma_{\overline{S}}$$

$$= i\dfrac{\sigma_{A\overline{S}}}{\sigma_A \sigma_{\overline{S}}} \sigma_A \qquad \text{multiplying above and below by } \sigma_A$$

$$= i r_{A\hat{A}} \sigma_A$$

The response to selection can be defined as the product of the standardised selection differential, i, the accuracy of the predicted breeding value, $r_{A\hat{A}}$, and the additive genetic standard deviation, σ_A, of the trait to be improved. The same procedure can be used for selection on progeny measurements or with repeated measurements on the individual, but in each case the response to selection is always equal to $i r_{A\hat{A}} \sigma_A$.

Information from relatives

Alternative selection procedures can be easily evaluated using the three parameters, i, $r_{A\hat{A}}$ and σ_A. If the proportion of selected animals is the same in each selection procedure, then alternative selection procedures can be evaluated solely on the basis of the accuracy of predicted breeding value. It is for this reason that Chapter 4 dealt with the variance of the predicted breeding value and prediction error variance, for estimation of the accuracy of the predicted breeding value.

For example, the responses to selection based on repeated measurements of potential sires, or measurements on progeny or measurements on full-sibs or on half-sibs were compared, given a phenotypic variance of 100, a heritability of 0.25 and standardised selection differential of one. For purposes of illustration, it has been assumed that the repeatability equals the heritability, and that there are no maternal or common environmental effects. Selected sires were mated to a random group of females, such that the selection differential on dams was zero and the overall standardised standard deviation was 0.5. Responses for the four selection procedures are illustrated in Fig. 5.2. The large range in the number of possible measurements was used to illustrate the response limit in each selection procedure. Obviously, in practical situations, it is unlikely that one animal will have more than five measurements or that the number of full-sibs, such as for pigs, will be greater than 15.

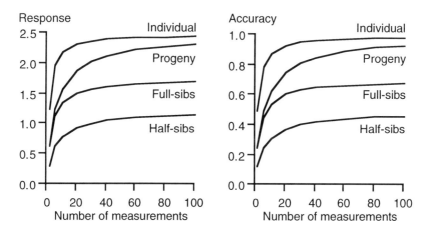

Fig. 5.2. Response and accuracy of selection for different numbers of measurements on the individual, and number of progeny, full- and half-sibs (assuming that $r_e = h^2$ and $c^2 = 0$)

As the number of measurements increases, the responses increase to an upper limit of 2.50, with selection on measurements on the animal or on progeny, but to limits of 1.77 $\left(= 2.5\sqrt{\frac{1}{2}}\right)$ and 1.25 $\left(= 2.5\sqrt{\frac{1}{4}}\right)$ with selection on full-sib and half-sib measurements, respectively. Similarly, as the number of measurements increases, the accuracy or breeding value prediction tends to one for repeated or progeny measurements, but to 0.71 $\left(=\sqrt{\frac{1}{2}}\right)$ and 0.5 for measurements on full-sibs and half-sibs, respectively. The graph of the accuracy of the predicted breeding value has exactly the same form as the responses graph (Fig. 5.2), as expected. The two graphs are shown to reinforce the point that

$$R = ir_{A\hat{A}}\sigma_A$$

The advantage of measurements on progeny is demonstrated, as the number of measurements on the individual or the number of full-sibs will not generally be greater than the number of progeny.

However, accuracies with more realistic parameters for the repeatability, 0.4, common environmental effect, $0.1\ \sigma_P^2$, the number of repeated measurements on the animal and the number of full-sibs are given in Fig. 5.3. The range of accuracies for the predicted breeding value of the individual with repeated measurements is similar to the range of accuracies with measurements on progeny and similarly with measurements on full-sibs and half-sibs. As the accuracy is a reflection of the response, then it is clearly important that unbiased estimates of the genetic and phenotypic parameters are required for appropriate evaluation of alternative selection strategies.

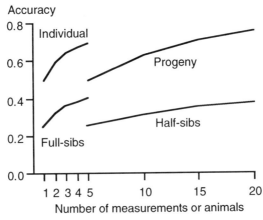

Fig. 5.3. Accuracy of selection for different numbers of measurements on the individual, and number of progeny, full- and half-sibs (assuming that $r_e = 0.4$ and $c^2 = 0.1\ \sigma_P^2$)

If a particular response is required, then the required number of measured progeny can be determined from the corresponding accuracy. For example, if the required improvement in carcass lean content is 40 g/kg, which has a heritability of 0.4 and a phenotypic standard deviation of 60 g/kg, and the selection proportions in boars and gilts are 0.10 and 0.20, then how many progeny with carcass information are required, per animal in the parental generation?

Firstly, the standardised selection differentials of boars and gilts are 1.755 and 1.400, such that the average standardised selection differential is 1.578 and the genetic standard deviation is 38 g/kg $\left(=\sqrt{0.4} \times 60 \text{ g/kg}\right)$.

Secondly, the required accuracy is 0.67 $(= \text{response}/i\sigma_A)$ and the number of progeny is seven, equal to

$$\frac{r_{A\hat{A}}^2}{1-r_{A\hat{A}}^2}\left(\frac{4-h^2}{h^2}\right).$$

For each of the animals being considered for selection, seven progeny with carcass composition information are required to achieve the desired response.

Chapter six
Selection Index Methodology

Several methods for predicting an animal's breeding value, given measurements on one trait on either the individual, sibs or progeny, have been described. In this chapter, methods developed in Chapters 4 and 5 will be extended to prediction of breeding values by combining measurements on the individual and its relatives for one trait or by combining measurements of several traits on the individual.

If the traits in the selection objective are the same traits as in the selection criterion, then one option may be to select on each trait in a sequential manner. Such selection is called "selection with independent culling levels", in that all animals with values for trait Y_1 greater than a threshold are selected for the second stage of selection, when those animals with trait Y_2 greater than a threshold are selected for the third stage, and so on. However, selection with independent culling levels is less efficient than using a selection criterion which combines measurements on all traits to predict genetic merit for the selection objective, using information on the genetic and phenotypic covariances. If the selection objective and selection criterion do not consist of the same traits, then selection with independent culling levels is not feasible.

Hazel (1943) developed a method of combining information from several sources, be it the same trait from different types of relatives or different traits measured on the animal, to predict the animal's genetic merit. The method for establishing a selection criterion for a given selection objective is referred to as selection index methodology.

Selection Objective and Selection Criterion

Selection objective

Assume that there are several traits which have to be improved, denoted by Y_1, Y_2, ..., Y_n, and that the traits have economic values of a_1, a_2, ..., a_n, respectively.

Selection index methodology

The economic value of a trait represents the additional economic return per marginal unit improvement in the trait. For example, the economic value of carcass lean content may be two economic units, assuming no change in food intake or growth rate.

Since the economic objective is to improve all traits, then the traits and their economic values are combined into a selection objective, where

$$\text{selection objective} = a_1 Y_1 + a_2 Y_2 + \ldots + a_n Y_n$$

The selection objective can be expressed in matrix notation as a'Y and is denoted by H.

For example, if growth rate and carcass lean content are to be improved and have economic values of 5 and 2, respectively, then the selection objective is

$$5 \times \text{growth rate} + 2 \times \text{carcass lean content}.$$

Selection criterion

The traits which are measured to predict the animal's breeding value are denoted $X_1, X_2, \ldots, X_m$. The measured traits are combined into an index on which the animals are selected. The selection index or selection criterion is as follows:

$$\text{selection criterion} = b_1 X_1 + b_2 X_2 + \ldots + b_m X_m$$

The selection criterion can be expressed in matrix notation as b'X and is denoted by I (for index).

For example, if growth rate and ultrasonic backfat depth are the traits measured, then animals are selected on the basis of

$$b_1 \times \text{growth rate} + b_2 \times \text{ultrasonic backfat depth}.$$

Traits in the selection criterion need not necessarily be the same traits as in the selection objective. For example, in one breeding programme, growth rate and carcass lean content are the traits to be improved, while growth rate and ultrasonic backfat depth are the traits measured for selection purposes. Similarly, the number of traits in the selection objective need not be the same as the number of traits in the selection criterion. In a second breeding programme, litter weight at weaning is to be improved, and the selection criterion consists of litter size and weight at birth.

The selection index method determines the selection criterion coefficients that maximise the response in the selection objective, H, with selection on the selection criterion, I. As several traits can be included in both the selection objective and the selection criterion, then information on the variances of the traits and on the relationships between the traits at the phenotypic and genetic levels is required. The information is in the form of three matrices:

P : the phenotypic variance–covariance matrix of traits in the selection criterion

G : the genetic covariance matrix between traits in the selection objective and the traits in the selection criterion

C : the genetic variance–covariance matrix of traits in the selection objective

Information on the necessary matrix algebra required for selection index methodology is included in the Appendix.

For example, if the selection objective is improvement in growth rate and carcass lean content, while the selection criterion consists of growth rate and ultrasonic backfat depth (bfat), then

$$P = \begin{bmatrix} var_P(growth) & cov_P(growth, bfat) \\ cov_P(growth, bfat) & var_P(bfat) \end{bmatrix}$$

or

$$P = \begin{bmatrix} \sigma^2_{growth} & r_P \sigma_{growth} \sigma_{bfat} \\ r_P \sigma_{growth} \sigma_{bfat} & \sigma^2_{bfat} \end{bmatrix}$$

$$G = \begin{bmatrix} var_A(growth) & cov_A(growth, lean) \\ cov_A(bfat, growth) & cov_A(bfat, lean) \end{bmatrix}$$

or

$$G = \begin{bmatrix} h^2_{growth} \sigma^2_{growth} & r_A h_{growth} h_{lean} \sigma_{growth} \sigma_{lean} \\ r_A h_{bfat} h_{growth} \sigma_{bfat} \sigma_{growth} & r_A h_{bfat} h_{lean} \sigma_{bfat} \sigma_{lean} \end{bmatrix}$$

$$C = \begin{bmatrix} var_A(growth) & cov_A(growth, lean) \\ cov_A(growth, lean) & var_A(lean) \end{bmatrix}$$

or

$$C = \begin{bmatrix} h^2_{growth} \sigma^2_{growth} & r_A h_{growth} h_{lean} \sigma_{growth} \sigma_{lean} \\ r_A h_{growth} h_{lean} \sigma_{growth} \sigma_{lean} & h^2_{lean} \sigma^2_{lean} \end{bmatrix}$$

where varp and covp are phenotypic variances and covariances, var$_A$ and cov$_A$ are additive genetic variances and covariances, rp and r$_A$ are the phenotypic and genetic correlations between traits and σ^2 is the phenotypic variance for a trait.

The P and C matrices are always symmetric. The G matrix will not generally be a symmetric matrix, as the number of traits measured will not always equal the number of traits to be improved and, secondly, the traits in the selection objective will not necessarily be the same traits as in the selection criterion.

Selection Criterion Coefficients

Derivation of the selection criterion coefficients for predicting genetic merit is more straightforward with matrix notation than it was in Chapter 4.

The variance of the selection criterion or the predicted genetic merit is

$$var(I) = var(b' X) = b' var(X) b = b' Pb$$

Similarly, the variance of the selection objective is

$$\text{var}(H) = \text{var}(a' Y) = a' \text{var}(Y)a = a' Ca$$

and, lastly, the covariance between the selection objective, H, and the selection criterion, I, is

$$\text{cov}(b' X, a' Y) = b' \text{cov}(X, Y)a = b' Ga$$

Given the P and G matrices and the economic values, a, of traits in the selection objective, the selection criterion coefficients can be defined as the coefficients that minimise the squared difference between the selection objective and the predicted genetic merit. The squared difference is

$$(H - I)^2 = (a' Y - b' X)^2$$
$$= a' \text{var}(Y)a - 2b' \text{cov}(X, Y)a + b' \text{var}(X)b$$
$$= a' Ca - 2b' Ga + b' Pb$$

Differentiation of the squared difference with respect to the selection criterion coefficients, b, gives

$$\frac{\partial (H - I)^2}{\partial b} = 2Pb - 2Ga$$

which, when equated to zero, results in

$$b = P^{-1} Ga$$

Two other derivations of the selection criterion coefficients are provided for completeness. The important point is that the selection criterion coefficients satisfy:

$$Pb = Ga$$

The selection criterion coefficients can also be defined as those which maximise the response in the selection objective. The response per standardised selection differential is

$$b_{HI} \sigma_I = \frac{\sigma_{IH}}{\sigma_I} = \frac{b' Ga}{\sqrt{b' Pb}}$$

Differentiation of the standardised response with respect to b, to identify the value of b that maximises the response, results in

$$\frac{\partial (b_{HI} \sigma_I)}{\partial b} = \frac{Ga}{\sqrt{b' Pb}} - b' Ga \frac{Pb}{\sqrt[3]{b' Pb}}$$
$$= \frac{1}{\sigma_I} \left[Ga - \frac{\sigma_{IH}}{\sigma_I^2} Pb \right]$$
$$= \frac{1}{\sigma_I} [Ga - Pb]$$

by setting $\frac{\sigma_{IH}}{\sigma_I^2} = b_{HI}$ to unity, so that the mean predicted breeding value is equal to the mean breeding value. Equating the partial differential to zero gives $Pb = Ga$. Therefore, the selection criterion coefficients are

$$b = P^{-1}Ga$$

A third derivation defines the selection criterion coefficients as those that maximise the correlation, or the accuracy between the selection criterion and the selection objective. However, as $r_{IH}\sigma_H = b_{HI}\sigma_I$ and the standard deviation of the selection objective, σ_H, is constant, then maximising the correlation is equivalent to maximising the response to selection per standardised selection differential, σ_I, since b_{HI} equals unity.

The predicted genetic merit for the selection objective, H, is b'X, and it is on the basis of the predicted genetic merit that animals are selected. In Chapter 4, the regression coefficient for the predicted breeding value on phenotypic measurement of an animal was $\frac{cov(A,P)}{var(P)}$. Similarly, when several traits are included in the selection objective and in the selection criterion, then the regression coefficient in the selection index method is $P^{-1}Ga$, where G equals cov(H,I), which is analogous to cov(A,P), and P equals var(I), which is analogous to var(P).

In Chapter 4, with selection on one trait, the regression of true breeding value on predicted breeding value was equal to one, such that an increase in the predicted breeding value corresponded to the same increase in true breeding value. With selection on several traits, using a selection criterion, the regression of H on I, of $\frac{cov(I,H)}{var(I)} = \frac{b'\,Ga}{b'\,Pb}$, is equal to unity, as expected.

In the remainder of this chapter, several parameters relating to genetic improvement of the selection objective with selection on the selection criterion are derived, to enable a series of different cases to be discussed, without having to derive further equations. The derivations of all equations regarding the selection index method are contained in a single chapter, rather than being spread throughout the text.

Responses to Selection

Selection for overall genetic merit in the selection objective, which consists of several traits, $Y_1, Y_2, ..., Y_n$, implies that the response in the overall selection objective will be the sum of the individual responses in each trait. The correlated response in trait Y_j, denoted by CR_j, to selection on the selection criterion is derived in the same manner as in Chapter 4:

$$CR_j = b_{jI}SD_I$$

where b_{jI} is the regression coefficient of trait Y_j on the selection criterion, I, and SD_I is the selection differential in the selection criterion

$$CR_j = i_I \frac{\text{cov}(Y_j, I)}{\sqrt{\text{var}(I)}}$$

where i_I is the standardised selection differential in the selection criterion, I.

The variance of the selection criterion has already been determined, as b'Pb. The covariance between the trait Y_j in the selection objective and the selection criterion is derived as

$$\text{cov}(Y_j, I) = \text{cov}\left(Y_j, \sum_{k=1}^{m} b_k X_k\right)$$

$$= \sum_{k=1}^{m} b_k \text{cov}(Y_j, X_k)$$

$$= b_1\sigma_{j1} + b_2\sigma_{j2} + \ldots + b_m\sigma_{jm}$$

$$= b'G_j \quad\quad \text{where } \sigma_{jm} \text{ is the genetic covariance between } Y_j \text{ and } X_m$$
where G_j is the jth column of the matrix G

The correlated response for a trait Y_j in the selection objective to selection on the selection criterion, I, is

$$CR_j = i_I \frac{b' G_j}{\sqrt{b' Pb}}$$

The correlated responses of traits, which are not included in the selection objective, can be calculated in a similar manner as for traits in the selection objective, by using the appropriate genetic covariances with the traits in the selection criterion.

The economic value of the response in the selection objective is the product of the responses for each trait in the selection objective multiplied by the corresponding economic values, CRa, where CR is the vector of correlated responses. The economic value of the response per standardised selection differential is

$$\frac{CRa}{i_I} = \frac{b' Ga}{\sqrt{b' Pb}}$$

The above equation is valid with any selection criterion coefficients, but if the selection criterion coefficients satisfy

$$Pb = Ga$$

then the economic value of the response per standardised selection differential is

$$\frac{CRa}{i_I} = \frac{b' Ga}{\sqrt{b' Pb}} = \sqrt{b' Pb}$$

which is the standard deviation of the selection criterion.

The accuracy of predicted genetic merit for any selection criterion is

$$r = \frac{\text{cov}(I, H)}{\sqrt{\text{var}(I)\,\text{var}(H)}}$$

$$= \frac{b'\,Ga}{\sqrt{(b'\,Pb)(a'\,Ca)}}$$

If the selection criterion coefficients satisfy $Pb = Ga$, then the accuracy is

$$r = \sqrt{\frac{b'\,Ga}{a'\,Ca}}$$

Scaled selection criterion

If a selection criterion, combining several traits, is used to identify animals of high genetic merit, then the between-animal variation in the selection criterion value will depend on the variance of the selection criterion. In certain situations, it would be useful if the intended variation for values of the selection criterion was fixed. For example, if the intended standard deviation for values of the selection criterion was 25, then it would be expected that the proportion of animals with a selection criterion value between -50 and 50 would be 0.95. Further, if the mean selection criterion value was set at 50, then the corresponding range of selection criterion values would be 0 to 100. Such a scaling of the selection criterion may be useful in commercial circumstances.

Scaling a selection criterion does not change the ranking of animals, such that the same animals are selected on both the unscaled and the scaled selection criteria. Both the accuracy of the scaled selection criterion and the predicted responses to selection using the scaled selection criterion are exactly the same as for the unscaled selection criterion.

If the intended variance of the scaled selection criterion is α^2, then the selection criterion coefficients of the scaled selection criterion are equal to

$$\tilde{b} = \sqrt{\frac{\alpha^2}{b'\,Pb}}\,b$$

since $\alpha^2 = \tilde{b}'\,P\tilde{b}$.

If the scaled selection criterion is denoted by $\tilde{I}$, then the accuracy of scaled selection criterion is

$$r = \frac{\text{cov}(\tilde{I}, H)}{\sqrt{\text{var}(\tilde{I})\,\text{var}(H)}}$$

$$= \frac{\tilde{b}'\,Ga}{\sqrt{\alpha^2 a'\,Ca}}$$

$$= \sqrt{\frac{b' Ga}{a' Ca}}$$

which is equal to the accuracy of the unscaled selection criterion.

The predicted response with selection on the scaled selection criterion is

$$\frac{CRa}{i_{\tilde{I}}} = \frac{\tilde{b}' Ga}{\sqrt{\tilde{b}' P \tilde{b}}}$$

$$= \frac{b' Ga}{\sqrt{b' Pb}}$$

which is the predicted response with the unscaled selection criterion.

Contribution of Traits in the Selection Objective

One method of assessing the contribution of a trait in the selection objective to the overall genetic merit is to determine the correlation between the trait and the selection criterion. If the correlation is low, then the response in the trait will be low, and will not contribute substantially to the response in overall genetic merit. If there are several traits to be considered for inclusion in the selection objective, then the selection objective should contain only the traits which will respond significantly to selection on the selection criterion. If too many traits are included in the selection objective, then this will be analogous to "trying to go in several directions at once, but going nowhere fast".

The correlation between a trait Y_j in the selection objective with the selection criterion, I, provides a method of identifying traits for inclusion in the selection objective and is

$$r_{YI}(j) = \frac{b' G_j}{\sqrt{b' Pb \, C_{jj}}}$$

where C_{jj} is the jth diagonal element of the C matrix, corresponding to the genetic variance of trait Y_j.

Contribution of Traits in the Selection Criterion

Similar to determining the contribution of a trait in the selection objective to the overall genetic merit, it is useful to determine the contribution of a trait in the selection criterion to the response in the selection objective. If a trait in the selection criterion does not significantly contribute information regarding traits in the selection objective, then there is little point of including the trait in the selection criterion, particularly if the trait is expensive or difficult to measure.

The contribution of a trait in the selection criterion to the response in the selection objective can be measured as the proportional reduction of the response in the selection objective, if the trait was excluded from the selection criterion. As discussed in Chapter 4, the accuracy of the predicted breeding value is proportional to the response to selection. Therefore, the contribution of a trait in the selection criterion to the selection objective is actually measured as the proportional reduction to the accuracy of the selection criterion if the trait was excluded from the selection criterion. The contribution of a trait in the selection criterion to the selection objective is

$$\frac{r_{IH}^*}{r_{IH}} = 1 - \sqrt{1 - \frac{b_j^2}{b' P b \, P_{jj}^{-1}}}$$

where r_{IH}^* is the accuracy of the selection criterion with the trait X_j omitted from the selection criterion and P_{jj}^{-1} is the jth diagonal element of the inverse of the P matrix (Cunningham, 1972).

Setting up the P, G and C matrices

Only the above formulae regarding the selection criterion coefficients, the accuracy of the selection criterion, the correlated responses to selection and the contribution of traits in the selection objective and in the selection criterion are required. The only difference between one selection procedure and another is in the elements of the P, G and C matrices. Calculation of the elements of the three matrices uses the equations derived in Chapters 4 and 5:

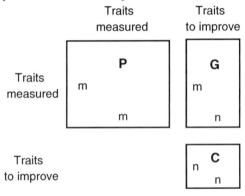

Fig. 6.1. Dimensions of P, G and C matrices

When determining the P, G and C matrices, i.e. "filling in the boxes", then the derivation of selection criteria and evaluation of selection procedures is straightforward (see Fig. 6.1). There are m traits measured and included in the selection criterion, so that P is a m × m matrix of phenotypic variances and covariances for the m traits in the selection criterion. Similarly, C is a n × n matrix of the genetic variances and covariances for the n traits included in the selection objective. Lastly, G is the m × n matrix of genetic covariances between the m traits in the selection criterion and the n traits in the selection objective.

The selection index methodology has been outlined in this chapter and there are examples of the use of the methodology in subsequent chapters.

As noted previously, information on the necessary matrix algebra required for selection index methodology is included in the Appendix. However, there are several computer packages which can be used for deriving selection criterion coefficients and the associated parameters, such as the accuracy of the selection criterion and the correlated responses to selection.

Chapter seven

Examples of Selection Objectives and Criteria

Now that formulae regarding the selection objective and criteria have been developed, examples of selection for several traits, using the selection index methodology, are presented. The first example is to simply equate the selection index formulae of Chapter 6 with those in Chapter 4, for selection on one trait.

Single measurement on an animal

From Chapter 4, an animal's breeding value can be predicted as

$$\hat{A} = h^2(P - P_{Pop})$$

with an accuracy equal to h, where h^2 is the heritability, P is the phenotypic measurement and P_{Pop} is the mean phenotype of the population. The response to selection is

$$R = ih^2\sigma_P$$

where $i\sigma_P$ is the selection differential.

In matrix notation, the P matrix equals σ_P^2 and the G and C matrices both equal σ_A^2. The economic value of the trait is arbitrarily set to one.

The selection criterion coefficients are

$$b = P^{-1}Ga = (\sigma_P^2)^{-1}\sigma_A^2 = h^2$$

and the accuracy of predicted genetic merit is

$$r_{IH} = \sqrt{\frac{b'\,Pb}{a'\,Ca}} = \sqrt{\frac{h^2\sigma_P^2 h^2}{\sigma_A^2}} = \sqrt{h^2} = h$$

with a response to selection of

$$CR = i_I \frac{b'\,G}{\sqrt{b'\,Pb}} = i_I \frac{h^2\sigma_A^2}{\sqrt{h^2\sigma_P^2 h^2}} = i_I \frac{\sigma_A^2}{\sigma_P} = i_I h^2 \sigma_P$$

The simple example illustrates how the selection index method can be used to determine the selection criterion coefficients, the accuracy of predicted genetic merit and the response to selection. Discussion on evaluating the contribution of traits in the selection objective and traits in the selection criterion requires more complex examples.

Measurements on an Individual and the Mean of its Sibs

The response to selection and the accuracy of predicted genetic merit for performance tested animals may be increased by incorporating measurements on an animal's sibs into the selection criterion. If X_1 is the animal's own performance test record and X_2 is the mean performance test record of its sibs, with the animal excluded, then the selection objective is X_1, as it is the animal's genetic merit which is to be predicted, and the selection criterion consists of X_1 and X_2.

It is arbitrary if the selection criterion consists of the individual and the mean of its sibs, with the individual included or excluded, or the individual's deviation from the family mean plus the family mean, as in each case the same total amount of information is used to predict the individual's breeding value. As illustrated in Question 16, at the end of the text, if one selection criterion is $b_1 \text{ind} + b_2 \overline{S}$, with the individual included in the sib mean, $\overline{S}$, and a second selection criterion is $b_3(\text{ind} - \overline{S}) + b_4 \overline{S}$, then the information used in the two criteria is the same, since $b_3(\text{ind} - \overline{S}) + b_4 \overline{S} = b_3 \text{ind} + (b_4 - b_3)\overline{S}$.

The P matrix is a 2 × 2 matrix of the variances of X_1 and X_2 and the covariance between X_1 and X_2, such that

$$P = \begin{bmatrix} \sigma_{X_1}^2 & \sigma_{X_1 X_2} \\ \sigma_{X_1 X_2} & \sigma_{X_2}^2 \end{bmatrix} = \begin{bmatrix} \sigma_P^2 & t\sigma_P^2 \\ t\sigma_P^2 & \left(\frac{1+(n-1)t}{n}\right)\sigma_P^2 \end{bmatrix} = \begin{bmatrix} 1 & t \\ t & \frac{1+(n-1)t}{n} \end{bmatrix} \sigma_P^2$$

The G matrix is a 2 × 1 matrix, as there are two traits in the selection criterion, X_1 and X_2, but only one trait in the selection objective, X_1, such that

$$G = \begin{bmatrix} \text{genetic var}(X_1) \\ \text{genetic cov}(X_1, X_2) \end{bmatrix} = \begin{bmatrix} \sigma_A^2 \\ r\sigma_A^2 \end{bmatrix} = \begin{bmatrix} h^2 \\ rh^2 \end{bmatrix} \sigma_P^2$$

For the P and G matrices, the covariance between an animal and the mean of its sibs is equal to the covariance between the animal and one of its sibs, $t\sigma_P^2$ and $r\sigma_A^2$, respectively.

Finally, the C matrix is $[\text{genetic var}(X_1)] = [\sigma_A^2] = [h^2]\sigma_P^2$.

Example

The food intake of pigs performance tested from 45 kg to 85 kg is to be reduced: this has a heritability of 0.2, a full-sib correlation of 0.25 and a phenotypic variance of 100 kg^2. Full-sib groups of six pigs are tested. Breeding values for each full-sib are required.

The above equations are used to determine the breeding value of each animal, calculated from the selection criterion consisting of the animal's food intake and the mean food intake of its five full-sibs. The P, G and C matrices are

$$\begin{bmatrix} 100 & 25 \\ 25 & 40 \end{bmatrix}, \begin{bmatrix} 20 \\ 10 \end{bmatrix} \text{ and } [20]$$

respectively, with a vector of economic weights, a, equal to minus one, since food intake is to be reduced. The selection criterion coefficients are

$$b = P^{-1}Ga = \frac{-1}{3375}\begin{bmatrix} 40 & -25 \\ -25 & 100 \end{bmatrix}\begin{bmatrix} 20 \\ 10 \end{bmatrix} = \begin{bmatrix} -0.163 \\ -0.148 \end{bmatrix}$$

The breeding value of an animal is predicted from

$$\hat{A} = -0.163(P - P_{Pop}) - 0.148(\bar{S} - P_{Pop})$$

where P is the food intake of the animal, $\bar{S}$ is the mean food intake of the animal's full-sibs and P_{Pop} is the population mean food intake.

The variance of the selection criterion is

$$b'Pb = [0.163 \quad 0.148]\begin{bmatrix} 100 & 25 \\ 25 & 40 \end{bmatrix}\begin{bmatrix} 0.163 \\ 0.148 \end{bmatrix} = 4.741$$

The accuracy of predicted genetic merit is

$$r_{IH} = \sqrt{\frac{b'Pb}{a'Ca}} = \sqrt{\frac{4.741}{20}} = 0.487$$

The correlated response in the animal's food intake

$$CR_{X_1} = i_I \frac{b'G_1}{\sqrt{b'Pb}} = -2.177 i_I$$

where $G_1 = \begin{bmatrix} 20 \\ 10 \end{bmatrix}$ and i_I is the standardised selection differential of the selection criterion. If an animal is selected on the basis of the selection criterion, then food intake is expected to reduce by 2.2 kg per standardised selection differential.

Although the mean food intake of the full-sibs is not included in the selection objective, the correlated response in mean food intake of the full-sibs can be determined, which requires the genetic covariance between mean food intake of the full-sibs with traits in the selection criterion. The genetic covariance between X_1 and X_2 is already known, and the genetic variance of the mean food intake for the full-sibs is

Examples of selection objectives and criteria

$$\left[\frac{1+(n-1)r}{n}\right]h^2\sigma_P^2$$

With $G_2 = \begin{bmatrix} 10 \\ 12 \end{bmatrix}$, the correlated response in mean food intake of the full-sibs is

$$CR_{X_2} = i_I \frac{b'G_2}{\sqrt{b'Pb}} = -1.564\, i_I$$

The mean food intake of the full-sibs also decreases, but to a lesser extent than the animal's food intake, when the response in full-sibs is treated as a correlated response to selection. Information from the full-sib mean is not a direct reflection of an animal's genetic merit, due to Mendelian sampling, such that the expected response of the animal is not totally reflected by the correlated response of the full-sib mean.

As it is expensive to measure food intake, it is sensible to determine the relative contribution to the selection criterion made by food intake of the animal and by the full-sib mean, using the formula

$$1 - \sqrt{1 - \frac{b_j^2}{b'Pb\, P_{jj}^{-1}}}$$

The diagonal elements of the inverse of P are 0.012 and 0.030, such that the relative contributions to the selection criterion by the animal and the full-sib mean are 0.27 and 0.08, respectively. If the selection criterion did not include the animal's food intake, then the accuracy of predicted genetic merit and the expected response to selection would be proportionately reduced by 0.27, but the proportional reduction would only be 0.08 if the mean food intake of full-sibs was not included in the selection criterion.

The contribution of a trait to the selection criterion can also be determined by directly comparing the accuracies of predicted genetic merit when the selection criteria includes or excludes the trait. For example, if selection was on the basis of the animal's own food intake, then the accuracy of the predicted breeding value would be the square root of the heritability, equal to 0.447. When mean food intake of full-sibs was excluded from the selection criterion, the proportional reduction in accuracy would be

$$1 - \frac{0.447}{0.487} = 0.08.$$

If all animals eligible for selection were performance tested and had food intake records, then there would be no extra cost incurred by including the food intake of full-sibs in the selection criterion. The response in food intake would be proportionally greater by 0.08 when the mean food intake of full-sibs was included in the prediction of an animal's breeding value. Given that the food intake information on full-sibs was available, anyway, then it would be inefficient to ignore the information when predicting breeding values.

The selection criterion consisted of the food intake of the animal and the mean of its full-sibs. Information on the mean food intake of the animal's half-sibs could be used to increase the accuracy of the predicted genetic merit. Similarly, information from parents and second-degree relatives could be included in the selection criterion, assuming that there were no substantial differences in the environments in which animals belonging to the parental and progeny generations were tested. Calculation of the selection criterion and related parameters would be similar to the example, except that the P and G matrices would be extended to accommodate the additional variances and covariances.

Measurement of Two Traits on the Individual

Inclusion of several traits measured on the individual into the selection objective and selection criterion is similar to incorporation of information from relatives into the selection criterion to predict the genetic merit of the individual for one trait. With two traits, X_1 and X_2, in the selection objective and criterion, the P and G matrices are

$$\begin{bmatrix} \sigma^2_{X_1} & r_P \sigma_{X_1} \sigma_{X_2} \\ r_P \sigma_{X_1} \sigma_{X_2} & \sigma^2_{X_2} \end{bmatrix} \text{ and } \begin{bmatrix} h_1^2 \sigma^2_{X_1} & r_A h_1 h_2 \sigma_{X_1} \sigma_{X_2} \\ r_A h_1 h_2 \sigma_{X_1} \sigma_{X_2} & h_2^2 \sigma^2_{X_2} \end{bmatrix}$$

where $\sigma^2_{X_1}$ and $\sigma^2_{X_2}$ are the phenotypic variances of traits X_1 and X_2.

The C matrix is equal to the G matrix, as the traits in the selection objective are the same as those in the selection criterion. The a matrix is $\begin{bmatrix} a_1 \\ a_2 \end{bmatrix}$.

Example

The number of eggs laid and the egg weight have economic values of 3 per egg and 60 per gram, with both traits to be improved in a breeding programme. The genetic and phenotypic parameters are as follows:

Trait		X_1	X_2	σ_P	Presentation	
Number of eggs	X_1	0.1	-0.1	30	h^2	r_P
Egg weight	X_2	-0.2	0.4	2	r_A	h^2

One style of presentation of parameters in the scientific literature is to present the heritabilities on the diagonal, with phenotypic correlations above the diagonal and genetic correlations below the diagonal.

The P and G matrices are $\begin{bmatrix} 900 & -6 \\ -6 & 4 \end{bmatrix}$ and $\begin{bmatrix} 90 & -2.4 \\ -2.4 & 1.6 \end{bmatrix}$, respectively.

The selection criterion is

$$0.29 \times \text{egg number} + 22.64 \times \text{egg weight}$$

and the accuracy of predicted genetic merit is 0.60. The correlated responses, the contribution of each trait to the selection objective and criterion, are as follows:

	Egg number	Egg weight (g)
Correlated response †	−0.62	0.78
Contribution to objective	−0.06	0.62
Contribution to criterion	0.02	0.91

† Correlated response per standardised selection differential of the selection criterion.

If selection was only on egg weight, then egg weight would increase by 0.80 g and egg number would reduce by 1.2 eggs, per unit selection differential, which would be economically worth 44.4. Selection on the full selection criterion is worth 45.2 per unit selection differential, such that exclusion of egg number from the selection criterion proportionally reduces the economic value of the response by 0.02; the contribution of egg number to the selection criterion.

Traits in the Selection Criterion but Not in the Selection Objective

It is not necessary for traits in the selection criterion to be included in the selection objective. For example, the selection objective may include growth rate and food intake, but the selection criterion may consist of only growth rate.

If trait Y is to be improved and the selection criterion consists of trait Y and a correlated trait X, measured on each animal, then the P, G and C matrices are

$$P = \begin{bmatrix} \sigma_Y^2 & r_P \sigma_X \sigma_Y \\ r_P \sigma_X \sigma_Y & \sigma_X^2 \end{bmatrix}, \quad G = \begin{bmatrix} h^2 \sigma_Y^2 \\ r_A h_Y h_X \sigma_Y \sigma_X \end{bmatrix}, \quad C = \begin{bmatrix} h_Y^2 \sigma_Y^2 \end{bmatrix}$$

The selection criterion coefficients are

$$b = P^{-1}G = \frac{1}{\sigma_Y^2 \sigma_X^2 (1 - r_P^2)} \begin{bmatrix} \sigma_X^2 & -r_P \sigma_Y \sigma_X \\ -r_P \sigma_Y \sigma_X & \sigma_Y^2 \end{bmatrix} \begin{bmatrix} h_Y^2 \sigma_Y^2 \\ r_A h_Y h_X \sigma_Y \sigma_X \end{bmatrix}$$

$$= \frac{1}{(1 - r_P^2)} \begin{bmatrix} h_Y^2 - r_A r_P h_Y h_X \\ (r_A h_Y h_X - r_P h_Y^2) \sigma_Y / \sigma_X \end{bmatrix}$$

The accuracy of the selection criterion is

$$\sqrt{\frac{h_Y^2 + r_A^2 h_X^2 - 2 r_P r_A h_X h_Y}{(1 - r_P^2)}}$$

When the selection objective is to improve trait Y, the effect of different values of the heritability of trait X, a correlated trait, and the genetic correlation between traits X and Y on the accuracy of the selection criterion, that includes both traits X and Y, is illustrated in Fig. 7.1. For example, given a heritability for trait Y of 0.3 and a phenotypic correlation between traits X and Y of 0.4, there is little advantage of including trait X in the selection criterion relative to direct selection on Y, for low values of the heritability of trait X when the genetic correlation is at least 0.5 or for high values of the heritability of trait X when the genetic correlation is 0.3. However, if the genetic correlation is only 0.1, then there is an advantage of including trait X in the selection criterion, irrespective of the heritability of trait X.

Clearly, reliable estimates of the genetic and phenotypic parameters are required when evaluating alternative selection strategies in breeding programmes.

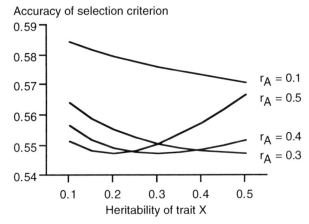

Fig. 7.1. Change in accuracy of selection criterion by inclusion of trait X in the selection criterion

However, there are instances in which inclusion of a correlated trait in the selection criterion may not necessarily improve the response to selection or the accuracy of predicted genetic merit. If the selection criterion coefficient for trait X is equated to zero, then

$$\frac{r_A h_Y h_X \sigma_Y \sigma_X}{h_Y^2 \sigma_Y^2} = \frac{r_P \sigma_Y \sigma_X}{\sigma_Y^2}$$

such that the genetic regression of X on Y equals the phenotypic regression of X on Y. In such a situation, trait X contributes no useful information regarding the genetic merit of trait Y, as the genetic correlation between traits X and Y is just a multiple of the environmental correlation:

$$r_A^2 = r_E^2 \frac{h_Y^2 \left(1 - h_X^2\right)}{h_X^2 \left(1 - h_Y^2\right)}$$

Example
The selection objective is to improve trait Y, and the selection criterion consists of traits X and Y. The genetic and phenotypic parameters are as follows:

	Y	X	σ_P
Y	0.10	0.60	10
X	0.30	0.40	10

with heritabilities on the diagonal, phenotypic correlations above the diagonal and genetic correlations below the diagonal.

The selection criterion is 0.1Y, with P and G matrices of

$$\begin{bmatrix} 100 & 60 \\ 60 & 100 \end{bmatrix} \text{ and } \begin{bmatrix} 10 \\ 6 \end{bmatrix}.$$

The accuracy of predicted genetic merit in Y is the square root of the heritability of Y, 0.316. Intuitively, it seems unusual that X contributes no additional information to the breeding value of Y, particularly as X is positively correlated with Y and has a higher heritability than Y. However, the genetic regression of X on Y is equal to the phenotypic regression, and so the selection criterion coefficient for X is zero.

Restricted Selection Objective

In certain situations it may be appropriate to constrain the correlated response of a particular trait to zero, while still maximising the rate of genetic improvement in the selection objective. For example, if one of the consequences of selection for carcass lean content is a reduction in reproductive performance, then it may be preferable to restrict the genetic change in reproductive performance to zero, but still improve carcass lean content. Similarly, genetic improvement in growth rate is generally associated with increased food intake, but it would be desirable to increase the growth rate without necessarily increasing food intake. The derivation of the selection criterion that results in zero genetic change in a particular trait requires a minor change to the existing formulae (Kempthorne and Nordskog, 1959; Cunningham *et al.*, 1970).

Two traits, Y_1 and Y_2, are included in the selection objective and the selection criterion consists of traits X_1 and X_2. The correlated response in trait Y_2 is to be restricted to zero.

The first step is to determine the selection criterion, as if there was no restriction on the genetic change in trait Y_2. The P, G, C and a matrices are

$$P = \begin{bmatrix} \text{var}_P(X_1) & \text{cov}_P(X_1, X_2) \\ \text{cov}_P(X_1, X_2) & \text{var}_P(X_2) \end{bmatrix}, \quad G = \begin{bmatrix} \text{cov}_A(X_1, Y_1) & \text{cov}_A(X_1, Y_2) \\ \text{cov}_A(X_2, Y_1) & \text{cov}_A(X_2, Y_2) \end{bmatrix}$$

$$C = \begin{bmatrix} \text{var}_A(Y_1) & \text{cov}_A(Y_1, Y_2) \\ \text{cov}_A(Y_1, Y_2) & \text{var}_A(Y_2) \end{bmatrix}, \quad a = \begin{bmatrix} a_{Y_1} \\ a_{Y_2} \end{bmatrix}$$

The selection criterion coefficients are $b = P^{-1}Ga$ and the selection criterion is

$$b_1(X_1 - \overline{X}_1) + b_2(X_2 - \overline{X}_2)$$

where $\overline{X}_1$ and $\overline{X}_2$ are the population mean values for traits X_1 and X_2.

The second step is to introduce a new variable, Y_3, into the selection criterion. There are now three traits in the selection criterion, so that the phenotypic variance–covariance matrix, of traits in the selection criterion, increases to a 3 × 3 matrix, which is denoted NP. The new phenotypic variance–covariance matrix, NP, is equal to the P matrix, with the column of G corresponding to trait Y_2 added as a column and as a row, and the remaining diagonal element of NP set to zero. The new phenotypic variance–covariance matrix, NP, is

$$NP = \begin{bmatrix} P & \begin{matrix} \text{cov}_A(X_1,Y_2) \\ \text{cov}_A(X_2,Y_2) \end{matrix} \\ \text{cov}_A(X_1,Y_2) \quad \text{cov}_A(X_2,Y_2) & 0 \end{bmatrix} = \begin{bmatrix} P & G_{Y_2} \\ G'_{Y_2} & 0 \end{bmatrix}$$

The genetic covariance matrix of traits in the selection criterion and in the selection objective must be changed, in accordance with the phenotypic variance–covariance matrix. There are now three traits in the selection criterion, but the selection objective, which consists of two traits, is unchanged, such that the new genetic covariance matrix, NG, is a 3 × 2 matrix. The new genetic covariance matrix is the G matrix with the addition of a row of zeros. The new genetic covariance matrix, NG, is

$$NG = \begin{bmatrix} G \\ \underline{0} \end{bmatrix}$$

where $\underline{0}$ represents a row of zeros.

As the response in trait Y_2 is restricted to zero, then the economic value of trait Y_2 is changed to zero, and the new vector of economic values, Na, is

$$Na = \begin{bmatrix} a_{Y_1} \\ 0 \end{bmatrix}$$

No change is made to the C matrix.

The coefficients of the new selection criterion, Nb, are calculated in the same manner as with an unrestricted selection criterion, as

$$Nb = (NP)^{-1} NG\, Na$$

The restricted selection criterion is

$$Nb_1(X_1 - \overline{X}_1) + Nb_2(X_2 - \overline{X}_2)$$

as the term Nb_3 is ignored.

The inferred economic value of trait Y_2 which results in no correlated reponse with an unconstrained selection criterion is $-Nb_3$.

Parameters relating to the selection objective and selection criterion, such as the accuracy of the predicted genetic merit and the correlated responses, are calculated with the formula for an unrestricted selection criterion, using matrices NP and NG.

Coefficients for the restricted selection criterion can also be derived from the equation (Brascamp, 1984)

$$b = P^{-1}\left[I - G_{Y_2}\left(G'_{Y_2} P^{-1} G_{Y_2}\right)^{-1} G'_{Y_2} P^{-1}\right] Ga = RGa$$

with G_{Y_2} equal to the column of G corresponding to trait Y_2.

The method of augmenting the P and G matrices to determine the selection criterion coefficients for the restricted selection criterion is equivalent to use of the Brascamp (1984) equation. The augmented matrices are:

$$\begin{bmatrix} P & G_{Y_2} \\ G'_{Y_2} & 0 \end{bmatrix} \begin{bmatrix} b \\ b_2 \end{bmatrix} = \begin{bmatrix} G \\ 0 \end{bmatrix} [a]$$

The first equation is as follows:
$$Pb + G_{Y_2} b_2 = Ga$$

After solving for b and multiplying by G'_{Y_2}, gives

$$G'_{Y_2} b = G'_{Y_2} P^{-1} Ga - G'_{Y_2} P^{-1} G_{Y_2} b_2$$

The second equation of the augmented matrices is as follows:

$$G'_{Y_2} b = 0$$

giving a solution for b_2 of

$$b_2 = \left(G'_{Y_2} P^{-1} G_{Y_2}\right)^{-1} G'_{Y_2} P^{-1} Ga$$

and incorporation into the first equation gives the Brascamp (1984) equation.

Example
The restricted selection objective is to improve growth rate, but to also constrain the genetic change in food intake to zero. Both growth rate and food intake are included in the selection criterion. The genetic and phenotypic parameters are

	Growth rate	Food intake	σ_P^2	Economic value
Growth rate (g/day)	0.45	0.50	100	1
Food intake (g/day)	0.60	0.30	200	−2

with heritabilities on the diagonal, phenotypic correlations above the diagonal and genetic correlations below the diagonal.

The P, G and a matrices are

$$P = \begin{bmatrix} 100 & 70.7 \\ 70.7 & 200 \end{bmatrix}, \quad G = \begin{bmatrix} 45 & 31.2 \\ 31.2 & 60 \end{bmatrix}, \quad a = \begin{bmatrix} 1 \\ -2 \end{bmatrix}$$

The G and C matrices are the same, as traits in the selection objective are equal to those in the selection criterion. The selection criterion coefficients are

$$b = P^{-1} Ga = \begin{bmatrix} 0.187 \\ -0.510 \end{bmatrix}$$

and animals would be selected on the basis of the index:

$$0.187 \times \text{growth rate} - 0.510 \times \text{food intake}$$

The correlated responses to selection are determined from

$$CR_j = i_1 \frac{b' G_j}{\sqrt{b' Pb}}$$

Growth rate and food intake would be reduced by −1.15 g/day and −3.82 g/day, respectively, per standardised selection differential. The undesirable reduction in growth rate indicates that use of a restricted selection index may be preferable.

As the response in food intake is to be restricted to zero, then the column of G corresponding to food intake is added to the new phenotypic variance–covariance matrix, NP, as a column and as a row. The remaining diagonal element of NP is set to zero. Therefore, NP is

$$NP = \begin{bmatrix} 100 & 70.7 & 31.2 \\ 70.7 & 200 & 60 \\ 31.2 & 60 & 0 \end{bmatrix}$$

A row of zeros is added to the new genetic covariance matrix, NG, which equals

$$NG = \begin{bmatrix} 45 & 31.2 \\ 31.2 & 60 \\ 0 & 0 \end{bmatrix}$$

and the economic value of food intake is set to zero, such that Na is

$$Na = \begin{bmatrix} 1 \\ 0 \end{bmatrix}$$

The coefficients of the new selection criterion, Nb, calculated using the new matrices, $Nb = (NP)^{-1} NG\, Na$, are

$$Nb = \begin{bmatrix} 0.358 \\ -0.186 \\ 0.718 \end{bmatrix}$$

The restricted selection criterion is now

$0.358 \times$ growth rate $-0.186 \times$ food intake

as the term $Nb_3 = 0.718$ is ignored in the selection criterion. The selection criterion now has relatively more emphasis on growth rate than on food intake, compared to the selection criterion when no constraints were imposed on the correlated responses.

If the selection criterion coefficients are calculated with the economic weight of food intake set at $-Nb_3$ and no constraints imposed on the correlated responses, then the correlated response in food intake will be zero. In the context of a restricted selection objective, the inferred economic weight of the trait, whose correlated response is restricted to zero, is equal to $-Nb_3$.

In a similar manner to the unrestricted selection objective, the correlated responses with selection on the restricted selection criterion are determined from

$$CR_j = i_1 \frac{Nb' NG_j}{\sqrt{Nb' NP\, Nb}}$$

and the responses in growth rate and food intake are 3.21 g/day and 0 g/day, respectively, per standardised selection differential. Selection on the restricted index will increase growth rate, without changing food intake.

Derivation of the selection criterion coefficients using the NP, NG and Na matrices may be considered to be more straightforward than using the Brascamp (1984) equation.

In the example,
$$R = \begin{bmatrix} 0.0124 & -0.0065 \\ -0.0065 & 0.0034 \end{bmatrix}, \text{ such that } b = \begin{bmatrix} 0.358 \\ -0.186 \end{bmatrix}$$

Desired Gains Selection Objective

The responses of specific traits in the selection objective may be required to equal predetermined values, while the rate of genetic response in other traits is maximised (Brascamp, 1984). The selection criterion coefficients are determined from the desired gains and the economic values of traits in the selection objective.

Two traits, Y_1 and Y_2, are included in the selection objective, and the selection criterion consists of traits X_1 and X_2. The correlated response in Y_2 is required to equal α times the response in Y_1, in numerical terms. For example, the desired response in daily food intake may be equal to twice the response in growth rate, or the desired response in carcass fat content may be equal to a multiple (presumably negative) of the response in carcass lean content. The coefficients of the desired gains selection criterion are determined from

$$b = S^{-1}RGa$$

where
$$S = I - \alpha P^{-1} G_{Y_2} \left(G'_{Y_2} P^{-1} G_{Y_2} \right)^{-1} G'_{Y_1}$$

with
$$R = I - G_{Y_2} \left(G'_{Y_2} P^{-1} G_{Y_2} \right)^{-1} G'_{Y_2} P^{-1}$$

and G_{Y_1} equal to the column of G corresponding to trait Y_1.

The restricted selection objective is clearly a special case of the desired gains selection objective, with α equal to zero in the equation for the S matrix, such that the response in trait Y_2 is equal to zero.

The inferred economic values of the traits, which correspond to the desired responses in the case of a unconstrained selection objective, a_D, are

$$\left(G' P^{-1} G \right)^{-1} G' b$$

Conversely, given the economic values, a_D, the responses with an unconstrained selection objective will be equal to those with the desired gains selection criterion.

Example

Using the parameters for growth rate and food intake in the restricted selection objective example, the correlated responses to an unconstrained selection objective are −1.15 g/day for growth rate and −3.82 g/day for food intake, and the economic value of the response is 6.48, per standardised selection differential (see Table 7.1). The accuracy of the corresponding selection criterion is 0.512.

When a restriction is imposed on the correlated response in food intake, the response in growth rate is 3.21 g/day, with an economic value of the response of 3.21, per standardised selection differential, and an accuracy of the restricted selection criterion of 0.254.

If the desired response in food intake was to equal 0.25 of the response in growth rate, then the correlated responses in growth rate and food intake would be 3.82 and 0.95 g/day, respectively, with the economic value of the response equal to 1.92, per standardised selection differential. The selection criterion coefficients would be 0.379 and −0.145. The accuracy of the selection criterion, given the desired gains selection objective, would be 0.198.

Table 7.1. Predicted responses to selection when the response in food intake is unconstrained or restricted and the responses to a desired gains selection objective

Selection objective	Response in		Economic value of response	Accuracy of selection criterion	Economic value of food intake
	Growth rate (g/day)	Food intake (g/day)			
Unconstrained	−1.15	−3.82	6.48	0.512	−2.0
Restricted	3.21	0	3.21	0.254	−0.718
Desired gains	3.82	0.95	1.92	0.198	−0.555[†]

† The economic value of food intake relative to the economic value of growth rate to achieve similar responses compared to the desired gains selection objective, given an unconstrained selection objective.

Constraining the responses to selection can substantially reduce the rate of gain in overall genetic merit, as indicated by both the reduction in the economic value of the response and in the accuracy of the selection criterion. In the example, the short term advantage of a reduction in food intake may be offset in the long term, by low food intake essentially imposing a constraint on genetic improvement in growth rate. The responses to the different selection objectives are summarised in Table 7.1.

Several aspects of selection index methodology have been discussed in this chapter, with examples to illustrate each point as they were developed. It is useful to have one comprehensive example to cover the points discussed.

Example from a Pig Breeding Programme

The selection objective is to increase carcass lean content (LEAN) and growth rate (GAIN) during a performance test from 30 kg to 85 kg, while daily food intake (FOOD) during the performance test is to be reduced. Although the food conversion ratio (FCR: food intake divided by weight gain) during the test is not included in the selection objective, the genetic change in food conversion ratio is to be monitored. The traits which can be included in the selection criterion are average daily gain (GAIN), average daily food intake (FOOD) and ultrasonic backfat depth (BFAT) measured at the end of the test. An animals' genetic merit is determined only from measurements on the animal. Incorporation of measurements from relatives will be discussed in the next section of this chapter.

There are four questions to be answered:

(1) What are the responses to selection, per unit selection differential of the selection criterion?
(2) What are the relative contributions of each trait in the selection objective and in the selection criterion?
(3) If the response in food intake is restricted to zero, how are the accuracy of predicted merit and the response in genetic merit changed?
(4) What is the efficiency of a selection criterion consisting only of growth rate and ultrasonic backfat depth?

The genetic and phenotypic parameters and the economic values of traits are as shown in Table 7.2, with heritabilities on the diagonal, phenotypic correlations above the diagonal and genetic correlations below the diagonal. The phenotypic correlations between GAIN, BFAT and FOOD with LEAN and FCR are not required, as the latter two traits are not included in the selection criterion.

Table 7.2. Genetic and phenotypic parameters for pig example

	GAIN (kg/day)	BFAT (mm)	FOOD (kg/day)	LEAN (dg/kg)	FCR (kg/kg)	σ_P	Economic value
GAIN	**0.46**	0.06	0.54			0.102	5
BFAT	-0.05	**0.31**	0.42			3.48	
FOOD	0.58	0.54	**0.34**			0.221	-5
LEAN	0.00	-0.60	-0.45	**0.45**		4.0	1
FCR	-0.60	0.55	0.30	0.40	**0.36**	0.209	

The economic values for growth rate and food intake are expressed relative to carcass lean content. For the purpose of the calculations, growth rate and food intake are expressed as kg/day rather than as g/day, while carcass lean content is expressed in decagrams, such that the phenotypic standard deviations of the five traits are of the same order of magnitude, to reduce rounding errors.

To set up the P, G and C matrices, the traits should be grouped according to those in the selection criterion and those in the selection objective, with var and cov indicating the phenotypic (subscript P) and additive genetic (subscript A) variances and covariances.

		Traits in selection criterion			Traits in selection objective		
		GAIN	BFAT	FOOD	GAIN	FOOD	LEAN
Traits in	GAIN	var_P	cov_P	cov_P	var_A	cov_A	cov_A
selection	BFAT	cov_P	var_P		cov_A	cov_A	
criterion	FOOD	cov_P	cov_P	**P**	cov_A	var_A	**G**
Traits in	GAIN				var_A	cov_A	cov_A
selection	FOOD				cov_A	var_A	
objective	LEAN				cov_A	cov_A	**C**

Calculation of the correlated responses for traits not in the selection objective requires the genetic variance–covariance matrix for traits in the selection criterion with traits not in the selection objective, $\mathbf{G}^*$.

		Traits NOT in the selection objective	
		BFAT	FCR
Traits in	GAIN	cov_A	cov_A
selection	BFAT	var_A	
criterion	FOOD	cov_A	$\mathbf{G}^*$

The elements of the matrices are as follows:

		Traits in selection criterion			Traits in selection objective		
		GAIN	BFAT	FOOD	GAIN	FOOD	LEAN
Traits in	GAIN	0.0104	0.0213	0.0122	0.0048	0.0052	0
selection	BFAT	0.0213	12.110	0.3230	-0.0067	0.1348	-3.119
criterion	FOOD	0.0122	0.3230	0.0488	0.0052	0.0166	-0.1556
Traits in	GAIN				0.0048	0.0052	0
selection	FOOD				0.0052	0.0166	-0.1556
objective	LEAN				0	-0.1556	7.200

		Traits NOT in the selection objective	
		BFAT	FCR
Traits in	GAIN	-0.0067	-0.0052
selection	BFAT	3.7542	0.1336
criterion	FOOD	0.1348	0.0048

Formulae for calculating the various parameters are as follows:

Selection criterion coefficients: $b = P^{-1}Ga$

Accuracy of predicted genetic merit: $r_{IH} = \sqrt{\dfrac{b'Pb}{a'Ca}}$

Correlated responses: $CR_j = i_I \dfrac{b'G_j}{\sqrt{b'Pb}}$

where G_j is the jth column of G or G*

Economic value of response: $CRa/i_I = \sqrt{b'Pb}$

Contribution of traits in the selection objective: $r_{YI}(j) = \dfrac{b'G_j}{\sqrt{b'Pb\, C_{jj}}}$

where G_j is the jth column of G

Contribution of traits in the selection criterion: $\dfrac{r_{IH}^*}{r_{IH}} = 1 - \sqrt{1 - \dfrac{b_j^2}{b'Pb\, P_{jj}^{-1}}}$

The questions are answered in order.

(1) What are the responses to selection, per unit selection differential of the selection criterion?

The responses in each trait are qualitatively as required by the selection objective, as growth rate and carcass lean content were increased, while food intake was reduced. Food conversion ratio was also reduced. Ultrasonic backfat depth was reduced in accordance with the increase in carcass lean content, as there was a large negative genetic correlation between the two traits:

Growth rate	Food intake	Carcass lean content	Backfat depth	Food conversion ratio
3 g/day	−55 g/day	10.1 g/kg	−1.1 mm	−0.06 kg/kg

The reduction in food intake may be acceptable in the short term, as a means of reducing food conversion ratio, but in the long term continued reduction in food intake may be unacceptable, if genetic improvement in the efficiency of lean growth is constrained or reproductive performance is impaired. Therefore, one option is to restrict the genetic change in food intake to zero.

(2) What are the relative contributions of each trait in the selection objective and in the selection criterion?

Correlations between the selection objective and growth rate, food intake or carcass lean content are 0.05, −0.43 and 0.38, respectively, which indicates that the response in economic merit is primarily due to the responses in food intake and carcass lean content. The economic value of the response in genetic merit is 1.306, per standardised selection differential, as the economic values of the responses in growth rate, food intake and carcass lean content are 0.017, 0.276 and 1.013, respectively.

If growth rate, ultrasonic backfat depth or food intake was not included in the selection criterion, then the accuracy of predicted genetic merit and, consequently, the response in genetic merit, would be proportionately reduced by 0.06, 0.13 or 0.16, respectively. Growth rate contributes less information regarding genetic merit for the selection objective than backfat depth and food intake.

(3) If the response in food intake is restricted to zero, how are the accuracy of predicted merit and the response in genetic merit changed?

Calculation of the restricted selection criterion requires the matrices NP and NG, and the vector of economic values, Na.

The NP matrix is $\begin{bmatrix} P & G_{FOOD} \\ G'_{FOOD} & 0 \end{bmatrix}$

where $G_{FOOD} = \begin{bmatrix} \text{cov}_A(\text{GAIN}, \text{FOOD}) \\ \text{cov}_A(\text{BFAT}, \text{FOOD}) \\ \text{var}_A(\text{FOOD}) \end{bmatrix} = \begin{bmatrix} 0.0052 \\ 0.1348 \\ 0.0166 \end{bmatrix}$

while the NG matrix is $\begin{bmatrix} G \\ 0 \; 0 \; 0 \end{bmatrix}$

and the Na vector is $\begin{bmatrix} 5 \\ 0 \\ 1 \end{bmatrix}$, since the economic value of food intake is set to zero.

With the restricted selection objective, responses to selection are as follows:

Growth rate	Food intake	Carcass lean content	Backfat depth	Food conversion ratio
35 g/day	0 g/day	7.3 g/kg	−0.9 mm	−0.07 kg/kg

Growth rate is substantially increased with a larger reduction in food conversion ratio compared to when the unrestricted selection criterion is used. Responses in carcass lean content and ultrasonic backfat depth are lower (7 versus 10 g/kg and 0.9 versus 1.1 mm, respectively) with the restricted selection objective, relative to the unrestricted objective.

The accuracies of predicted genetic merit with and without the restriction on food intake are 0.34 and 0.43, respectively. The corresponding responses in genetic merit, per standardised selection differential, with and without the restriction on food intake, are 0.907 and 1.306. In both cases, carcass lean content proportionately accounts for 0.80 of the economic value of the response. The reduction in economic value is primarily due to the smaller response in carcass lean content with the restricted selection objective, rather than the reduced economic value from a restriction on food intake, which is compensated for by an increase in growth rate. The effect of the restriction on genetic change in food intake is proportionally to reduce the response in economic merit by 0.40, largely due to a reduction in the response of carcass lean content.

Examples of selection objectives and criteria

The substantial decrease in response due to the restriction on genetic change in food intake may not be a viable alternative to the reduction in food intake with the unrestricted selection objective. Another alternative would be to exclude food intake from the selection criterion, which may reduce the magnitude of the response in food intake, and as food intake is an expensive trait to measure, the reduction in costs may be beneficial.

(4) What is the efficiency of a selection criterion consisting only of growth rate and ultrasonic backfat depth?

The new P, G, C and a matrices are sub-matrices of the matrices calculated for the original selection objective and criterion.

With the new selection criterion, the responses to selection are as follows:

Growth rate	Food intake	Carcass lean content	Backfat depth	Food conversion ratio
4 g/day	−37 g/day	9.0 g/kg	−1.1 mm	−0.04 kg/kg

The responses in growth rate, carcass lean content and ultrasonic backfat depth are similar to when food intake was included in the selection criterion. The magnitude of the responses in food intake and food conversion ratio are lower compared to the original selection criterion and the economic value of the response (1.103 versus 1.306) is proportionately reduced by 0.16. Although the efficiency of the selection criterion to predict genetic merit is reduced, the decrease in the magnitude of genetic change in food intake, from 55 g/day to 37 g/day, may be acceptable in both the short and the long term.

A summary of the selection criterion coefficients, accuracies and responses for each selection objective is given in Table 7.3.

Table 7.3. Selection criterion parameters and responses when the response in food intake is unconstrained or restricted and when food intake is not recorded

	Selection criterion coefficients			Accuracy of selection	Economic value of response
	Growth rate	Backfat depth	Food intake		
Selection objective	5.23	−0.21	−4.26	0.43	1.31
Food restricted	7.29	−0.17	−0.90	0.34	0.91
Food not recorded	0.46	−0.32	—	0.37	1.10

	Growth rate (g/day)	Food intake (g/day)	Carcass lean (g/kg)	Backfat depth (mm)	Food conversion ratio (kg/kg)
Selection objective	3	−55	10.1	−1.1	−0.06
Food restricted	35	0	7.3	−0.9	−0.07
Food not recorded	4	−37	9.0	−1.1	−0.04

Selection on Several Traits with Information from Relatives

In Chapters 4 and 5, the prediction of breeding values for single or for several traits was based on information on the individual, with either one measurement or repeated measurements, or information from one group of relatives. In this section, measurements on several traits, both on the individual and on its relatives, are combined to predict the genetic merit of the individual.

The equations to predict genetic merit of the individual for the different combinations of information are summarised in Table 7.4, with P equal to the individual's measurement and $\overline{P}$ equal to the mean measurement on its relatives. For simplicity, a population mean phenotype of zero, has been assumed.

Table 7.4. Formulae for predicting genetic merit of the individual from information on the individual and relatives

Selection objective	Traits in the selection criterion		
	Individual	Relatives	Individual and relatives
Single trait	$I = h^2 P$	$I = \dfrac{nrh^2}{1+(n-1)t}\overline{P}$	$I = b_1 P + b_2 \overline{P}$
Several traits	$I = b_1 P_1 + b_2 P_2$	$I = b_1 \overline{P}_1 + b_2 \overline{P}_2$	$I = \Sigma b_i P_i + \Sigma b_j \overline{P}_j$

The methodology to predict genetic merit, the accuracy of the prediction and the response to selection is the same in each case. The phenotypic variance–covariance matrix of traits in the selection criterion, P, the genetic covariance of traits in the selection criterion with traits in the selection objective, G, the genetic variance–covariance matrix of traits in the selection objective, C, and the economic value of each trait in the selection objective, a, are required. The selection criterion coefficients are $b = P^{-1}Ga$. The only difference between the different circumstances is in calculation of the elements of the P, G and C matrices.

The example from pig breeding is again used to illustrate the combination of information on several traits measured on both the individual and its relatives. The selection objective is to increase growth rate and carcass lean content. Each performance tested animal has its growth rate measured during the test and its ultrasonic backfat depth measured at the end of the test. Measurements are available on the individual and the mean of its n full-sibs, such that there are four traits in the selection criterion: growth rate and backfat depth measured on the individual and the full-sib mean measurement for growth rate and ultrasonic backfat depth, with the individual excluded.

The P, G and C matrices are formed, firstly, by grouping the traits according to those included in the selection criterion, those included in the selection objective and then, secondly, by calculating the elements of the matrices ("filling in the boxes").

Examples of selection objectives and criteria

In the example, there is a degree of overlap between the four traits in the selection objective with the two traits in the selection objective. One approach would be to code the traits in the selection criterion as X_1, X_2, X_3 and X_4 and the traits in the selection objective as Y_1 and Y_2, which may result in unnecessary confusion, as there are only three traits in total: growth rate, ultrasonic backfat depth and carcass lean content. Therefore, the selection criterion consists of the growth rate and backfat depth of the individual, denoted by growth and backfat, and the full-sib mean growth rate and ultrasonic backfat depth measurements, denoted by $\overline{\text{growth}}$ and $\overline{\text{backfat}}$, respectively.

In the equations for elements of the matrices, the three traits have been coded as follows: 1, growth rate; 2, ultrasonic backfat depth; 3, carcass lean content; with subscripts P or A denoting a phenotypic or an additive genetic variance and similarly for covariances.

The P matrix, for traits in the selection criterion, is

	Growth	Backfat	$\overline{\text{Growth}}$	$\overline{\text{Backfat}}$
Growth	σ_{P1}^2	σ_{P12}	$t_1\sigma_{P1}^2$	$\frac{1}{2}\sigma_{A12} + \sigma_{C12}$
Backfat		σ_{P2}^2	$\frac{1}{2}\sigma_{A12} + \sigma_{C12}$	$t_2\sigma_{P2}^2$
$\overline{\text{Growth}}$			$\left(t_1 + \frac{1-t_1}{n}\right)\sigma_{P1}^2$	$\frac{\sigma_{P12}}{n} + \frac{n-1}{n}\left(\frac{1}{2}\sigma_{A12} + \sigma_{C12}\right)$
$\overline{\text{Backfat}}$				$\left(t_2 + \frac{1-t_2}{n}\right)\sigma_{P2}^2$

with t_1 and t_2 equal to the correlation between full-sibs for traits 1 and 2.

The G matrix, for traits in the selection criterion with those in the selection objective, is

	Growth	Carcass lean content
Growth	σ_{A1}^2	σ_{A13}
Backfat	σ_{A21}	σ_{A23}
$\overline{\text{Growth}}$	$\frac{1}{2}\sigma_{A1}^2$	$\frac{1}{2}\sigma_{A13}$
$\overline{\text{Backfat}}$	$\frac{1}{2}\sigma_{A21}$	$\frac{1}{2}\sigma_{A23}$

and the C matrix, for traits in the selection objective, is

	Growth	Carcass lean content
Growth	σ_{A1}^2	σ_{A13}
Carcass lean content	σ_{A31}	σ_{A3}^2

For clarity of presentation, only the upper triangular part of the P matrix has been given. The equation for the phenotypic covariance between the full-sib means for growth rate and ultrasonic backfat depth is:

$$\text{cov}(\overline{\text{growth}}, \overline{\text{backfat}}) = \text{cov}\left[\frac{1}{n}\sum_{i=1}^{n}\text{growth}_i, \frac{1}{n}\sum_{j=1}^{n}\text{backfat}_j\right]$$

$$= \frac{1}{n^2}\sum_{i=1}^{n}\left[\sum_{j=1}^{n}\text{cov}(\text{growth}_i, \text{backfat}_j)\right]$$

$$= \frac{1}{n^2}n\left[\sigma_{P12} + (n-1)\left(\frac{1}{2}\sigma_{A12} + \sigma_{C12}\right)\right]$$

The double summation is essentially equal to adding up all the elements of a square, with n rows and n columns. A row represents the covariance between growth rate for animal i with backfat depth of its (n−1) full-sibs and itself, while a column represents the covariance between backfat depth for animal j with growth rate of its (n−1) full-sibs and itself:

Sib	1	2	3	4
1	σ_{P12}	$\frac{1}{2}\sigma_{A12} + \sigma_{C12}$	$\frac{1}{2}\sigma_{A12} + \sigma_{C12}$	$\frac{1}{2}\sigma_{A12} + \sigma_{C12}$
2	$\frac{1}{2}\sigma_{A12} + \sigma_{C12}$	σ_{P12}	$\frac{1}{2}\sigma_{A12} + \sigma_{C12}$	$\frac{1}{2}\sigma_{A12} + \sigma_{C12}$
3	$\frac{1}{2}\sigma_{A12} + \sigma_{C12}$	$\frac{1}{2}\sigma_{A12} + \sigma_{C12}$	σ_{P12}	$\frac{1}{2}\sigma_{A12} + \sigma_{C12}$
4	$\frac{1}{2}\sigma_{A12} + \sigma_{C12}$	$\frac{1}{2}\sigma_{A12} + \sigma_{C12}$	$\frac{1}{2}\sigma_{A12} + \sigma_{C12}$	σ_{P12}

There is no difference between adding up the elements of the square by rows or by columns, as each row or column will contain the same elements. For example, with four full-sibs, the row corresponding to the second animal contains the phenotypic covariance between growth and backfat depth for itself, σ_{P12}, and one half of the genetic plus common environmental covariance between growth and backfat depth, $\frac{1}{2}\sigma_{A12} + \sigma_{C12}$, for each its three full-sibs. Row two is

$$\sigma_{P12} + (4-1)\left(\frac{1}{2}\sigma_{A12} + \sigma_{C12}\right)$$

and as there are four rows, then the sum of the elements is

$$4\left[\sigma_{P12} + (4-1)\left(\frac{1}{2}\sigma_{A12} + \sigma_{C12}\right)\right]$$

which has to be divided by $\frac{1}{4^2}$ for calculation of the covariance.

The covariance between the full-sib mean for growth rate and the full-sib mean for ultrasonic backfat depth is

$$\frac{1}{n}\left[\sigma_{P12} + (n-1)\left(\frac{1}{2}\sigma_{A12} + \sigma_{C12}\right)\right]$$

Examples of selection objectives and criteria 89

The same genetic and phenotypic parameters are used as in the example from the previous section. The common environmental effects for growth rate and ultrasonic backfat depth account for $0.1\,\sigma_P^2$, such that the correlation between full-sibs is one half of the heritability plus 0.1. The common environmental covariance between growth rate and ultrasonic backfat depth is zero.

Changes in accuracy of the selection criterion, the selection criterion coefficients and the correlated responses to selection, as the number of full-sibs increase, are presented in Table 7.5 and illustrated graphically in Fig. 7.2.

Table 7.5. Accuracy of the selection criterion, the selection criterion coefficients and the correlated responses to selection according to the number of full-sib measurements in the selection criterion

			\multicolumn{5}{c}{Number of full-sibs (excluding the individual animal)}				
		Animal	1	2	3	4	5
Accuracy	r_{IH}	0.35	0.36	0.37	0.37	0.38	0.38
Selection criterion coefficients	growth	2.84	2.64	2.59	2.54	2.50	2.48
	$\overline{\text{growth}}$		0.32	0.47	0.55	0.60	0.62
	backfat	−0.26	−0.25	−0.24	−0.23	−0.22	−0.22
	$\overline{\text{backfat}}$		−0.07	−0.11	−0.14	−0.16	−0.18
Correlated responses	growth rate	16.2	15.8	15.6	15.4	15.2	15.1
	carcass lean content	8.7	9.0	9.2	9.3	9.4	9.4

As the number of full-sibs with measurements increases, the accuracy of predicted genetic merit increases (Fig. 7.2). Genetic improvement in carcass lean content increases with an increasing number of full sibs measured, while the response in growth rate decreases. The accuracy increases to a limit of 0.40, and the more full-sib information; the better. The selection criterion coefficients for growth rate, and for backfat depth, of the individual and full-sib mean change in a complementary manner as the number of full-sibs increases. The selection criterion coefficients tend to limits of 2.36 and 0.64 for growth rate on the individual and the full-sib mean, and −0.18 and −0.31 for backfat depth. Although the accuracy increases as the number of measured full-sib increases, if performance test facilities are limited, then it may be more important to increase the number of families tested and selected, such that variation in inbreeding is not increased, rather than reduce the number of families tested and selected, by increasing the number of full-sibs per family.

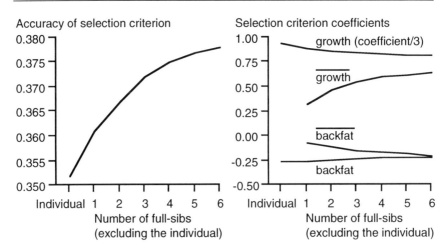

Fig. 7.2. Accuracy of selection criterion and selection criterion coefficients with increasing numbers of full-sibs

Economic Values

The economic value of a trait represents the additional economic return per marginal unit improvement in the trait. Economic values can be expressed relative to an animal, the producer or the national interest: this has been examined by Brascamp *et al.* (1985) and Smith *et al.* (1986). Weller (1994) comprehensively discusses economic aspects of animal breeding. Therefore, there is no need for a detailed derivation of economic values within this text.

For illustrative purposes, economic values for a pig breeding context are determined for continuous and discrete traits. One method of estimating economic values is to change the trait of interest by a unit amount, relative to the mean of the trait, and to determine the change in the cost of production, assuming that all other traits remain fixed at their current levels.

For example, assume that the mean growth rate and food intake of pigs grown from 30 to 85 kg are 850 g/day and 1.9 kg/day, respectively. Food costs 150 economic units per tonne and there is an overhead cost of 0.20 economic units per day.

The economic weight for growth rate can be calculated as the reduction in costs between a pig growing at 860 g/day compared to one growing at 850 g/day. The economic gain by increasing growth rate by 10 g/day is

$$\left[\frac{(85-30)}{0.850} - \frac{(85-30)}{0.860}\right] \times (1.9 \times 0.150 + 0.20) = 0.365$$

such that the economic value for growth rate is 0.036 economic units.

Similarly, the economic value of daily food intake can be derived from the increased food costs of a pig eating 10 g/day more than a pig eating 1.9 kg/day, with no change in growth rate:

$$\frac{(85-30)}{0.850} \times 0.150 \times (1.91-1.90) = 0.097$$

and the economic value for daily food intake is 0.010 economic units.

Carcass shape is graded for processing purposes, with the carcass price depending on the grade. The distribution of carcasses and the carcass price per kg for each grade is given in Table 7.6.

Table 7.6. Distribution of carcasses and the carcass price per kg by grade

Grade	1	2	3	4
Distribution of carcasses	0.50	0.30	0.15	0.05
Carcass price per kg	105	100	95	85

Under the assumption of an underlying normal distribution for carcass grade, the thresholds of t_1, t_2 and t_3 between the four carcass grades, with corresponding ordinates of z_1, z_2 and z_3, are shown in Figure 7.3.

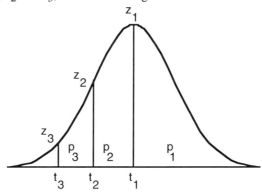

Fig. 7.3. Thresholds and ordinates corresponding to proportions, assuming a normal distribution

If the proportion of carcasses in each grade are p_1, p_2, p_3 and p_4, with corresponding carcass prices of c_1, c_2, c_3 and c_4, then the marginal cost of a unit change in the underlying scale (Meijering, 1986) is

$$(c_1-c_2)z_1 + (c_2-c_3)z_2 + (c_3-c_4)z_3$$

For carcass grade, the economic value is

$$(105-100) \times 0.40 + (100-95) \times 0.28 + (95-85) \times 0.10 = 4.4$$

If the selection objective of the breeding programme is to increase growth rate and carcass grade, but to reduce daily food intake, then the economic value for daily food intake is negative, to reflect the intended reduction, as shown in Table 7.7. Economic values for the three traits differ substantially, but not when expressed in terms of phenotypic standard deviations. Note that the derived economic values are for illustrative purposes only.

Table 7.7. Examples of economic values for pig production traits

	Economic value	Standard deviation (s.d.)	Economic value per s.d.
Growth rate (g/day)	0.036	100	3.6
Daily food intake (g/day)	−0.010	220	−2.1
Carcass grade	4.4	1	4.4

Changes in the economic values of traits in the selection objective can have a substantial effect on the accuracy of a selection criterion (Smith, 1983). Reductions in the accuracy will depend on changes in the economic value of traits with both relatively high economic values and relatively high heritabilities. If the vectors of the correct and incorrect economic values are a and $\tilde{a}$, then the accuracy of the selection criterion with incorrect economic values for traits relative to the accuracy with the correct economic values is:

$$\frac{\tilde{b}' Pb}{\sqrt{(\tilde{b}' P\tilde{b})(b' Pb)}}$$

where $b = P^{-1}Ga$ and $\tilde{b} = P^{-1}G\tilde{a}$.

For example, traits X and Y are included in both the selection objective and selection criterion, with heritabilities equal to 0.3 and 0.2, respectively, with equal phenotypic and genetic correlations. The correct economic values of traits X and Y are two and one, and the economic value of trait X is changed. The relative accuracy of the selection criterion with incorrect economic weights is illustrated in Fig. 7.4. When the economic value of trait X is less than the correct value, there is a substantial reduction in accuracy, while there is little loss of accuracy with a positive change to the economic value of trait X.

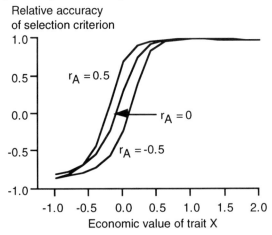

Fig. 7.4. Relative accuracy of the selection criterion when changes are made to economic value of trait Y

Chapter eight

Factors Affecting the Rate of Genetic Improvement

Errors in Genetic and Phenotypic Parameters

Differences between estimated and population genetic and phenotypic parameters will have an effect on the predicted response to selection. Given the population parameters, the formula for the actual responses of traits in the selection objective for a given selection criterion, was discussed in Chapter 6, and is

$$R = \frac{\text{cov}(I, H)}{\text{var}(I)} SD_I = i \frac{b' G}{\sqrt{b' Pb}}$$

The predicted responses given the estimated selection criterion coefficients, $\hat{b}$, derived from the estimated $\hat{G}$ and $\hat{P}$ matrices are

$$\hat{R} = i \frac{\hat{b}' \hat{G}}{\sqrt{\hat{b}' \hat{P} \hat{b}}}$$

but the actual responses given the estimated selection criterion coefficients are

$$R^* = i \frac{\hat{b}' G}{\sqrt{\hat{b}' P \hat{b}}}$$

If the estimated genetic and phenotypic parameters are not equal to the population parameters, then the response will be less than the maximum response. The loss in efficiency due to the difference between the estimated and population parameters (Sales and Hill, 1976a) is

$$1 - \frac{R^*}{R}$$

One trait in the selection objective and criterion

When the selection objective is to improve one trait and the selection criterion consists of measurements on the individual and its relatives, then errors in estimates of the intra-class correlation and the heritability will have little effect on the response (Sales and Hill, 1976a). For example, if the selection criterion is based on measurements on an individual and the mean of its four half-sibs, then the loss in efficiency increases as the deviation between the population and estimated intra-class correlation increases. However, for heritabilities between 0.05 and 0.8, the loss in efficiency is less than 0.02 when the deviation between the population and estimated half-sib correlation is 0.1.

Two traits in the selection criterion

If the selection objective is to improve trait Y and there is a correlated trait X, included in the selection criterion, then the selection criterion coefficients are

$$b = P^{-1}G = \frac{1}{(1-r_P^2)} \begin{bmatrix} h_Y^2 - r_A r_P h_Y h_X \\ (r_A h_Y h_X - r_P h_Y^2) \sigma_Y / \sigma_X \end{bmatrix}$$

The need for reliable estimates of the genetic and phenotypic parameters for evaluation of alternative selection strategies was discussed in Chapter 7, in the section on "Traits in the selection criterion, but not in the selection objective". The advantage of including a second trait in the selection criterion was largely dependent on the genetic correlation between traits.

However, in the special case when the population parameters satisfy $r_A h_X = r_P h_Y$, then the response to selection, R, will be a multiple of the heritability for trait Y. If the estimated genetic correlation is $r_A + \alpha$, then the actual response to selection with the selection criterion coefficients determined from the parameter estimates, R^*, is proportional to

$$\frac{h_Y^4}{\sqrt{h_Y^4 + \Delta}}, \quad \text{where } \Delta = \frac{\alpha^2 h_X^2 h_Y^2}{1-r_P^2}$$

The predicted response to selection given the selection criterion coefficients, $\hat{R}$, is proportional to $\sqrt{h_Y^4 + \Delta}$, with a predicted gain in efficiency by incorporating trait X in the selection criterion of

$$\frac{\hat{R}}{R} = \frac{\sqrt{h_Y^4 + \Delta}}{h_Y^2}$$

The actual response to selection on the estimated selection criterion relative to selection on the selection criterion derived from the population parameters is

$$\frac{R^*}{R} = \frac{h_Y^2}{\sqrt{h_Y^4 + \Delta}}$$

which is exactly the reciprocal of the predicted gain in efficiency. Therefore, the actual loss in efficiency and the predicted gain in efficiency increase as the difference between the estimated genetic correlation and the population genetic correlation, α, increases, and as the absolute value of the phenotypic correlation increases. When the estimated parameters differ from the population's parameters and $r_A h_X = r_P h_Y$ (Sales and Hill, 1976b), then the predicted gain in efficiency from including trait X in the selection criterion is equal to the actual loss in efficiency.

Example

If the difference between the estimated genetic correlation and the population genetic correlation is 0.4 in magnitude, and the phenotypic correlation is equal to ±0.4, then, as $\left(h_X^2 - h_Y^2\right)$ increases, the actual loss in efficiency increases proportionately by up to 0.40, when h_Y^2 is greater than 0.1 (Fig. 8.1). As the heritability of trait Y increases, the actual loss in efficiency decreases when X is included in the selection criterion.

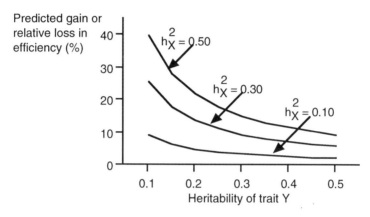

Fig. 8.1. Predicted gain or relative loss in efficiency by including trait X in the selection objective to improve trait Y, when $r_A h_X = r_p h_Y$, relative to selection on trait Y only

Modification of Parameter Estimates

Negative variance component estimates can be obtained from ANOVA analyses, such that the corresponding heritability estimate will be negative. For example, given the following liveweights of progeny from four sires, in Table 8.1:

Table 8.1. Liveweights of progeny from four sires

Sire A	Sire B	Sire C	Sire D
12	13	15	14
14	14	15	12
15	15	16	16
18	16	18	14

the corresponding analysis of variance is given in Table 8.2:

Table 8.2. Analysis of variance for progeny liveweights

Source of variation	DF	Sum of squares	Mean squares	Expectation of mean squares
Between-sires	3	8.69	2.90	$\sigma_e^2 + 4\sigma_s^2$
Within-sires	12	37.75	3.15	σ_e^2

The estimated between-sire variance component is –0.062 and the corresponding heritability estimate is –0.08. Clearly, prediction of breeding values and responses to selection, given a negative heritability estimate, will be of no value.

For selection on several traits, valid genetic and phenotypic parameters are required to determine selection criterion coefficients, as otherwise predicted responses to selection will be unreliable. Estimates of the heritability, genetic and phenotypic correlations can be examined for inconsistent values, such as heritabilities outside the range of 0 to 1 and correlations outside the range of –1 to +1. One method of determining whether the phenotypic and genetic variance–covariance matrices, P and G, are consistent with expectations is to examine the eigenvalues of the matrix $P^{-1}G$ (Hayes and Hill, 1980).

Eigenvalues and eigenvectors

The following section on eigenvalues and eigenvectors and the discussion on bending is included in the text for completeness. Knowledge of matrix algebra, other than covered in the Appendix, is required. However, the section can be omitted, if necessary.

For a symmetric matrix, A, there is a diagonal matrix, D, and a square matrix, V, such that

$$V'AV = D \quad \text{and} \quad VV' = I$$

The diagonal elements of the D matrix are eigenvalues, while the columns of the V matrix are eigenvectors (Rao, 1973), which are the solutions to the equation

$$|A - \lambda I| = 0$$

and the eigenvectors are determined by solving $(A - \lambda I)x = 0$.

If $A = \begin{bmatrix} 9 & 3 \\ 3 & 4 \end{bmatrix}$, then the eigenvalues are 2.59 and 10.40, with corresponding eigenvectors of $\begin{bmatrix} -0.424 \\ 0.906 \end{bmatrix}$ and $\begin{bmatrix} 0.906 \\ 0.424 \end{bmatrix}$. As a check, the sum of the eigenvalues is the sum of the diagonal terms of the A matrix.

If the phenotypic and genetic variance–covariance matrices, P and G, are estimated in a multivariate analysis, then the corresponding analysis of variance table, for s sires each with n half-sib progeny is given in Table 8.3, noting that

$$P = [B + (n-1)W]/n \quad \text{and} \quad G = 4(B-W)/n$$

Table 8.3. Formula for analysis of variance table

Source of (co)variation	DF	Mean squares	Expectation of mean squares
Between-sires	s−1	B	$P - \frac{1}{4}G + \frac{1}{4}nG$
Within-sires	s(n−1)	W	$P - \frac{1}{4}G$

The eigenvalues of the matrix $P^{-1}G$ are determined from $|P^{-1}G - \lambda I| = 0$ or $|G - \lambda P| = 0$. If all eigenvalues of $P^{-1}G$ are positive and less than one, then the heritabilities, genetic and phenotypic correlations will be within the appropriate parameter spaces of (0,1) and (−1, 1), respectively, as will the partial correlations. If the two conditions for the eigenvalues of $P^{-1}G$ are satisfied, then selection index coefficients and predicted responses to selection determined from the P and G matrices may be reliable, given the precision of the estimated P and G matrices. Determining whether the eigenvalues of the $P^{-1}G$ matrix are between 0 and 1 is the multivariate equivalent of checking whether the estimated heritability of a trait lies between 0 and 1.

Given ANOVA estimates of the P and G matrices, the probability of $P^{-1}G$ having a negative eigenvalue increases as the number of traits increases and as the number of progeny per sire decreases, as illustrated in Fig. 8.2, with data on the half-sib progeny of 20 sires and the eigenvalues of $P^{-1}G$ equal to 0.25 (Hill and Thompson, 1978).

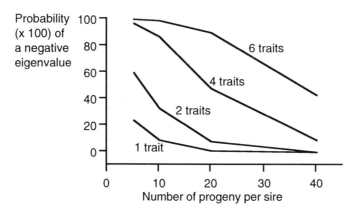

Fig. 8.2. The probability of a negative eigenvalue for the $P^{-1}G$ matrix

The mean of the eigenvalues of the estimated $P^{-1}G$ matrix is unbiased, but the larger eigenvalues are positively biased, while the smaller eigenvalues are negatively biased. The magnitude of the bias increases as the number of traits increases and as the sample size decreases. If the spread of eigenvalues of the estimated $P^{-1}G$ matrix was reduced, then the reliability of the predicted accuracy of a selection criterion may be improved. One method of reducing the spread of eigenvalues is called "bending" (Hayes and Hill, 1981).

In particular, if some eigenvalues of estimated $P^{-1}G$ matrix are either less than zero or greater than one, then the "bending" procedure can be used to ensure that eigenvalues of the modified $P^{-1}G$ matrix are non-negative and less than one. Rather than modifying the P and G matrices directly, the eigenvalues of the B and W matrices are "bent". The eigenvalues of the $W^{-1}B$ matrix are equivalent to eigenvalues of the $P^{-1}G$ matrix, as $\lambda = \dfrac{4(v-1)}{(v-1+n)}$, where v are the eigenvalues of $W^{-1}B$, obtained by solving $|B - vW| = 0$, since the P and G matrices are linear combinations of the B and W matrices. The degrees of freedom for within-sires are generally greater than the degrees of freedom for between-sires, such that the W matrix is more precisely estimated than the B matrix. Only the eigenvalues of the B matrix are "bent" and the modified B matrix, B*, is

$$B^* = (1-\gamma)B + \gamma \bar{v} W$$

where $0 < \gamma < 1$ and $\bar{v}$ is the mean eigenvalue of $W^{-1}B$.

The modified P^* and G^* matrices are

$$G^* = 4\left[(1-\gamma)B - (1-\gamma\bar{v})W\right]/n$$
$$P^* = \left[(1-\gamma)B + (n-1+\gamma\bar{v})W\right]/n$$

The optimal value of γ depends on the eigenvalues and the number of traits. One approach is to determine the smallest value of γ that results in the smallest eigenvalue of $P^{*^{-1}}G^*$ being equal to zero.

Generation Interval

The response to selection per generation is

$$\frac{i_m + i_f}{2} h^2 \sigma_P$$

where i_m and i_f are the standardised selection differentials of selected males and females, respectively, for selection on an animal's single measurement. The response to selection per unit time is

$$\frac{i_m + i_f}{L_m + L_f} h^2 \sigma_P$$

where L_m and L_f are the respective generation intervals for selected males and females. The generation interval is defined as the average age of parents when their offspring are born, where the offspring are parents of the next generation. The length of the generation interval must be considered when evaluating alternative selection strategies.

Example

Two alternative breeding programmes are to be evaluated. The first programme selects animals of the basis of n repeated measurements on an individual, with a repeatability of r_e and heritability of h^2, and a generation interval, L, of one year. The repeated measurements are made before mating, such that the generation interval is not increased. In the second breeding programme, animals are progeny tested, with each of the np half-sib progeny measured once. The time delay in obtaining the progeny measurements increases the generation interval to two years. How many progeny are required, per animal being considered for selection, to obtain the same rate of genetic improvement as selection on the animal's measurements?

The responses per generation can be compared using the generation interval and accuracies of predicted merit. The accuracies with n repeated measurements on the individual or one measurement on each of np half-sib progeny are

$$\sqrt{\frac{n}{n\frac{r_e}{h^2}+\frac{1-r_e}{h^2}}} \quad \text{and} \quad \sqrt{\frac{np}{np+\frac{4-h^2}{h^2}}}$$

The number of progeny required per animal being considered for selection is

$$n_P = \frac{nL^2(4-h^2)}{1-r_e + n(r_e - L^2 h^2)}$$

If there are two measurements on the individual for a trait with a heritability of 0.1 and repeatability of 0.2, 0.3 or 0.4, then at least 78, 63 or 52 progeny would be required to obtain a similar annual response compared to selection on the individual's mean measurement. It may not be possible to select the same proportion of animals with individual testing as with progeny testing, due to a limit on the testing facilities for large numbers of progeny, such that the selection proportion with progeny testing may be lower than with repeated measurements.

Reduction in Variance (Bulmer Effect)

Selection on a trait reduces both the phenotypic and genetic variances, such that prediction of long-term responses based on the response to one generation of selection will tend to overestimate the long-term response (Bulmer, 1971; Robertson, 1977).

Let σ_P^2 and σ_A^2 be the phenotypic and additive genetic variances in the population before selection. The same proportion of males and females are selected as parents, with selection on the basis of an animal's one measurement. The phenotypic variance of the selected parents is reduced by $k\sigma_P^2$, where $k = i(i-x)$, with i and x, determined from normal distribution tables, corresponding to the mean of the selected animals, in phenotypic standard deviation units, and the threshold value, above which animals are selected. The phenotypic variance of the selected parents is:

$$(1-k)\sigma_P^2$$

The genetic variance of the selected parents and their progeny are:

$$(1-h^2k)\sigma_A^2 \quad \text{and} \quad \left(1-\frac{1}{2}h^2k\right)\sigma_A^2$$

The additive genetic variance within full-sib families is not changed by selection and remains at $\frac{1}{2}\sigma_A^2$, but the between full-sib family variance is reduced to $\frac{1}{2}(1-h^2k)\sigma_A^2$, such that the additive genetic variance of full-sib progeny is:

$$\left(1-\frac{1}{2}h^2k\right)\sigma_A^2$$

For example, with a selection proportion in males and females of 0.20, i and x equal 1.40 and 0.842, and the changes in variance and response per generation are as shown in Table 8.4.

Table 8.4. Changes in genetic and phenotypic parameters with selection

	\multicolumn{5}{c}{Generations of selection}				
	0	1	2	3	4
σ_P^2	100	90.2	88.1	87.6	87.5
σ_A^2	50	40.2	38.1	37.6	37.5
$\sigma_{Disequilibrium}^2$	0	−9.8	−11.9	−12.4	−12.5
Heritability	0.5	0.446	0.432	0.429	0.428
Response		7	5.93	5.68	5.62

There is a substantial reduction in the response per generation after the first round of selection, due to the reduction in additive genetic variance and heritability. The environmental variance is assumed to be constant between generations, such that the phenotypic variance after t generations of selection is equal to the additive genetic variance plus the environmental variance:

$$\sigma_{P(t+1)}^2 = \sigma_{A(t+1)}^2 + \sigma_E^2$$

The additive genetic variance after t generations of selection depends on the proportion of animals selected in generation t and the disequilibrium variance at generation t:

$$\sigma_{A(t+1)}^2 = \left[1-\frac{1}{2}h_{(t)}^2 k\right]\sigma_{A(t)}^2 - \frac{1}{2}\sigma_{Disequilibrium(t)}^2$$

The disequilibrium variance is the difference between the additive genetic variance after t generations of selection and prior to selection. The disequilibrium variance is due to genes segregating together as a result of selection:

$$\sigma^2_{Disequilibrium} = \sigma^2_{A(t)} - \sigma^2_{A(0)}$$

The additive genetic variance after t generations of selection is

$$\sigma^2_{A(t+1)} = \frac{1}{2}\left[\sigma^2_{A(t)}\left(1 - h^2_{(t)}k\right) + \sigma^2_{A(0)}\right]$$

The reduction in genetic variance due to selection has several consequences. Predicted long term responses to selection will be overestimated, as shown in Table 8.4, when calculated using a heritability estimate from an unselected population. Secondly, predicted responses with selection based on between-family or within-family deviations will be incorrect, due to the redistribution of the genetic variance between and within families. Thirdly, if progeny tested sires are highly selected, then the value of additional information from sibs may be less than predicted, due to the reduction in genetic variance.

Inbreeding

Inbreeding is the mating of animals that have ancestors in common, such that at a particular locus their progeny may be homozygous for an allele, which belonged to one ancestor. The inbreeding coefficient, F, for progeny of two individuals, is half the genetic relationship between the individuals, which was discussed in Chapter 5. For example, the inbreeding coefficients for progeny of full-sibs and progeny of half-sibs are 0.25 and 0.125, respectively. Evaluation of alternative breeding programmes should take account of both the rate of genetic improvement and the rate of inbreeding, due to the detrimental effects of inbreeding depression.

Inbreeding depression

In an inbred population, if there are n loci affecting a trait, with allele frequencies p_i and dominance deviations d_i, then the reduction in the mean phenotypic performance of

$$2F \sum_{i=1}^{n} p_i (1 - p_i) d_i$$

relative to the outbred population, due to the level of inbreeding is called inbreeding depression. Calculation of inbreeding depression assumes that there is no interaction between the n loci affecting the trait. The effect of a 1% increase in inbreeding coefficient of the individual and of the dam has been a reduced litter size of 0.013 and 0.023 in pigs (Hill and Webb, 1982), a reduced weaning weight in cattle of 0.44 and 0.30 kg (Burrow, 1993), and a reduced milk yield of 25 kg (Miglior et al., 1995).

Rate of inbreeding

The rate of inbreeding in a population, ΔF, is the difference in the average inbreeding coefficient for the population at generation t+1, F_{t+1}, and at generation t, F_t, relative to the difference between complete inbreeding, when the inbreeding coefficient is one, and the level of inbreeding at generation t:

$$\Delta F = \frac{F_{t+1} - F_t}{1 - F_t}$$

The predicted rate of inbreeding in an "idealised" population, which is a population of size N, consisting of equal numbers of males and females, selected and mated at random, and with random family size, is

$$\Delta F = \frac{1}{2N}$$

In a non-"ideal" population, the rate of inbreeding is predicted from

$$\Delta F = \frac{1}{2N_e}$$

where N_e is the effective population size, corresponding to the population size required to achieve the observed rate of inbreeding in an "idealised" population. When the numbers of breeding males and females are unequal, the effective population size can be approximated by

$$N_e \approx \frac{4N_m N_f}{N_m + N_f}$$

where N_m and N_f are the number of breeding males and females in the population, and the rate of inbreeding is

$$\Delta F \approx \frac{1}{8N_m} + \frac{1}{8N_f}$$

As the variance of the family size increases, the effective population size decreases, such that the rate of inbreeding increases.

Incorporation of variation in family size into the formula for the rate of inbreeding (Hill, 1979) assumes that variation in family size has no genetic component. In a selected population, variation in family size will have a genetic component, as animals of high genetic merit will have more selected progeny than animals of low genetic merit, such that the rate of inbreeding will be underestimated. Several studies have developed methods to predict the rate of inbreeding in selected populations (Woolliams *et al.*, 1993; Wray *et al.*, 1994).

Variance of response and inbreeding

The rate of genetic improvement is a function of the selection differential, the accuracy of selection and the genetic variance, with a successive reduction in the genetic variance as the number of generations of selection increases, i.e. the Bulmer effect. The variance of the rate of genetic improvement, ΔG, also depends on the rate of inbreeding, ΔF:

$$\text{var}(\Delta G) = 2\Delta F \sigma_{A\infty}^2 \left(1 - r_\infty^2 k\right)$$

where $\sigma^2_{A\infty}$ and r_∞ are the asymptotic values for the genetic variance and accuracy of selection after several generations of selection (Meuwissen and Woolliams, 1994). An increase in the rate of inbreeding will be accompanied by increased variation in the rate of genetic improvement, such that the realised rate of genetic improvement may differ substantially from the predicted rate of genetic improvement. Simultaneous maximisation of the rate of genetic improvement and minimisation of the difference between the predicted and observed rates of genetic improvement will be difficult to achieve, as both parameters depend on the proportion of animals selected to be parents of the next generation, which determines, to an extent, the rate of inbreeding. However, several methods have been proposed which markedly decrease the rate of inbreeding, and therefore the variance of genetic response, without substantially reducing the rate of genetic improvement.

Methods to reduce the rate of inbreeding

One method of reducing the rate of inbreeding, while maintaining the rate of genetic improvement, is to increase the population size. Several approaches have been examined to constrain the rate of inbreeding, without increasing the size of the population. The approaches can be categorised according to family structure, selection on biased predicted breeding value, and the combination of predicted breeding value and inbreeding coefficient.

Family structure

The rate of inbreeding increases with the variance of family size, but the effective population size can be maximised by selecting animals within families, to ensure that a female from each full-sib family replaces her dam and a male from each half-sib family replaces his sire (Hill *et al.*, 1996).

The rate of genetic improvement for within-family selection is lower than with mass selection, such that the advantage of reducing the rate of inbreeding is offset by the reduced rate of genetic improvement. Increasing the number of half-sib families, by reducing the number of full-sibs in a family, would reduce the rate of inbreeding without reducing the rate of genetic improvement.

For example, in a dairy cattle breeding programme, one procedure is for immature follicles to be collected from a donor cow with the ova matured *in vitro* and ova fertilised using semen from different bulls (Woolliams, 1989). Fertilised ova would be transferred to recipient cows. Given 36 donor cows mated to four bulls with eight ova fertilised per cow, the predicted rate of genetic improvement with nine full-sib families of size eight, for each bull, was the same as with 18 full-sib families of four offspring per bull, but the predicted rate of inbreeding, as measured by the effective population size, was lower.

Selection on biased predicted breeding values

Inclusion of information from relatives for prediction of breeding values increases the correlation between relatives' predicted breeding values and also the probability of selection for relatives of animals with high genetic merit. One method of reducing the probability of selecting relatives is to increase the value of the heritability used to determine the selection criterion coefficients, such that the selection criterion weighting on the mean measurement of the animal's relatives is reduced (Grundy *et al.*, 1994).

For example, if the selection criterion is based on measurements of an individual and its six full-sibs, and the heritability of 0.25 is incremented by 0.01, 0.025 or 0.05, then the selection criterion coefficient for the mean measurement of the full-sibs changes from being 1.80 times the coefficient for the individual's measurement to 1.77, 1.70 and 1.62, respectively, such that there is relatively less emphasis on the full-sib information.

The rate of inbreeding can be reduced substantially, with a only marginal reduction in the rate of genetic improvement, with the latter being due to the use of a non-optimal selection criterion. The increment to the heritability for calculation of the selection criterion coefficients depends on the heritability and the required rate of inbreeding.

Predicted breeding value and inbreeding coefficient

When predicting breeding values, rather than altering the balance of information from the animal and its relatives, the breeding values could be adjusted for inbreeding directly, to reduce the rate of inbreeding. The animal's predicted breeding value could be adjusted for either its inbreeding coefficient, F (Toro and Perez-Encisco, 1990), or for the average genetic relationship between the animal and other selected animals, $\bar{r}$ (Brisbane and Gibson, 1995), such that the adjusted breeding value would be either

$$\hat{A} - kF \quad \text{or} \quad \hat{A} - k\bar{r}$$

respectively. Examination of the predicted rates of genetic improvement and inbreeding for a series of k values is required to determine the appropriate k value.

Adjustment of an animal's breeding value for its inbreeding coefficient is straightforward once the value of k has been determined. In contrast, adjustment of the individual's breeding value for the average genetic relationship between the individual and other selected animals requires knowledge of the identity of the selected animals, which can only be determined after their breeding values have been adjusted. Therefore, breeding values of all animals are initially calculated and animals with the highest predicted breeding values are allocated to the selected group, with their breeding values subsequently adjusted for the genetic relationships. Animals in the selected group with lower adjusted breeding values than animals in the unselected group are replaced, and the breeding values are adjusted for the new genetic relationships. The process is repeated until no new animals are selected.

Generally, large reductions in the variance of the rate of genetic improvement can be obtained, without substantially reducing the rate of genetic improvement, by either incrementing the heritability when predicting breeding values or by adjusting the breeding values for the animal's inbreeding coefficient or for its genetic relationship with other selected animals. However, the optimum value for the adjustment factor has to be determined empirically, as no formal method of determining the optimum adjustment factor has been proposed.

Chapter nine

Performance Testing, Progeny Testing and MOET

Several aspects of predicting breeding values and responses to selection have been examined and a series of examples have been used to illustrate each new point. Measurements have been made on the individual and its relatives for one or several traits to enable prediction of the individual's breeding value, with all of the animals belonging to the same generation. However, performance testing and progeny testing are complementary methods of providing information for breeding value prediction. Similarly, measurements on animals from previous generations can be incorporated with measurements on an animal and its relatives for prediction of breeding values. In dairy cattle breeding, bulls are evaluated using information from their dam's generation, their sibs and their progeny, such that information is available from several generations. In this chapter, the combination of performance testing and progeny testing is examined, with a sheep breeding example, and incorporation of information from several generations is discussed with regard to dairy cattle breeding.

Optimising Performance Testing and Progeny Testing

The selection objective of a particular sheep breeder is to improve carcass lean content in the crossbred progeny of purebred rams. A total of five rams are to be selected from 100 performance tested rams. Growth rate and ultrasonic backfat depth are measured on all rams, and are combined into an index by the sheep breeder. The index of growth rate and ultrasonic backfat depth has a heritability of 0.25 and a genetic correlation with carcass lean content of crossbred progeny of 0.6, which has a heritability of 0.4 and a phenotypic standard deviation of 30 g/kg. The sheep breeder could select the rams on the basis of the performance test index, with a generation interval of one year.

However, the sheep breeder also owns a flock of crossbred ewes, with carcass information available on a total of 500 progeny. The sheep breeder could also select the rams on the basis of a progeny test, which has a generation interval of two years. The question posed by the sheep breeder is: What is the optimal combination of performance testing and progeny testing, to identify the five rams for selection?

The performance test index is denoted as trait 1, with carcass lean content denoted as trait 2. The genetic parameters are $h_1^2 = 0.25$, $h_2^2 = 0.4$ and $r_A = 0.6$.

One option is to select the five rams on the basis of the performance test. For purposes of the example, only the selection of rams is considered, such that the annual rate of genetic improvement is equal to half the genetic merit of the selected rams divided by the generation interval. The annual rate of response is

$$R_{Performance} = \frac{0.5 b_{21} SD_1}{L_{Performance}}$$

where b_{21} is the regression of predicted breeding value for carcass lean content on the performance test index and SD_1 is the selection differential for the performance test index. As five rams are selected from 100, then the selection proportion is 0.05 and the standardised selection differential is 2.063. The regression coefficient is $\frac{r_A h_1 h_2 \sigma_1 \sigma_2}{\sigma_1^2}$. Therefore, with performance testing only, the annual response in carcass lean content is 5.9 g/kg.

Alternatively, the five rams could be selected on the basis of the progeny test, with an annual rate of response of

$$R_{Progeny} = \frac{0.5 b_{A\bar{P}} SD_{\bar{P}}}{L_{Progeny}}$$

where $b_{A\bar{P}}$ is the regression of predicted breeding value for carcass lean content on mean progeny carcass lean content, equal to $\frac{2nh^2}{4+(n-1)h^2}$, and $SD_{\bar{P}}$ is the selection differential based on mean progeny carcass lean content. Since there are 500 progeny from 100 rams, then the number of progeny per ram is five and the regression coefficient is 0.714. The value of $SD_{\bar{P}}$ is $i_2 \sqrt{\frac{1+(n-1)t}{n}} \sigma_2$, equal to $0.529 i_2 \sigma_2$, since all progeny are half-sibs and $t = \frac{1}{4} h_2^2$. Therefore, with progeny testing only, the annual response in carcass lean content is 5.8 g/kg.

There is little difference between the annual rates of improvement with selection on either performance test or on progeny test, in the example.

The third option is to select a proportion of rams on the basis of performance test, progeny test those rams and finally select the required five rams on the basis of performance test and progeny test (Cunningham, 1975). The combination of performance test and progeny test makes full use of the performance test facilities; and progeny testing of a subset of the 100 rams, rather than all 100 rams, will increase the number of progeny per ram, to increase the accuracy of selection. If the genetic correlation between the performance test index and carcass lean content is substantially lower than one, then the rams of high predicted genetic merit for the performance test index may not have high predicted genetic merit for carcass lean content. Therefore, the advantage of increasing the accuracy of the predicted genetic merit for carcass lean content, by increasing the number of progeny per sire, may be offset by the disadvantage of excluding rams of high genetic merit for carcass lean content from the progeny test.

The response, with both performance testing and progeny testing, is

$$R_{Combine} = i_1 r_A h_1 h_2 \sigma_2 + \frac{i_2 h_2^2 \left(1 - r_A^2 h_1^2 k\right) \sigma_2}{\sqrt{h_2^2 \left(1 - r_A^2 h_1^2 k\right) + \frac{4 - h_2^2}{n}}}$$

The annual rate of response is $0.25 R_{Combine}$ to take account of the generation interval of two years, due to progeny testing, and assuming that the rate of genetic improvement is equal to one half of the genetic merit of selected rams. The first part of the equation, $i_1 r_A h_1 h_2$, is the correlated response in trait 2, carcass lean content, given selection on trait 1, the performance test index. The second part of the equation is the response with selection on the basis of progeny test, with the variance in trait 2 reduced by $\left(1 - r_A^2 h_1^2 k\right)$, to take account of the prior selection on trait 1.

The terms i_1 and i_2 reflect the proportion of rams selected on the performance test index and on the basis of the progeny test, respectively, with $k = i_1(i_1 - x_1)$. When all rams are progeny tested, then $i_1 = 0$ and the equation reduces to the expected response with selection on mean progeny measurement, as discussed in Chapter 5.

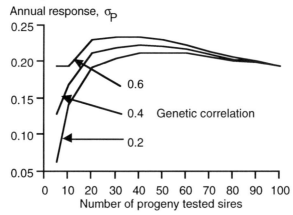

Fig. 9.1. Annual rates of genetic improvement in carcass lean content for different numbers of rams progeny tested

For the example, the optimum number of rams to progeny test was determined empirically, by varying the number of rams selected on the performance test index. The annual rates of genetic improvement in carcass lean content for different numbers of rams progeny tested, with several values of the genetic correlation between the index of performance test traits and carcass lean content, are illustrated in Fig. 9.1.

The annual response in carcass lean content of 7.1 g/kg, from progeny testing 40 selected performance tested sires, is proportionately 0.20 higher than the annual response with either performance testing or progeny testing alone.

The example illustrates the combination of performance and progeny testing for one generation of selection. However, the reduction in genetic variance would have to be accounted for in the evaluation of the three testing procedures over several generations of selection.

Progeny Testing and MOET (Multiple Ovulation and Embryo Transfer)

In conventional dairy cattle breeding programmes, the predicted genetic merit of bulls is based on progeny performance, which has a high accuracy, provided that sufficient daughters are recorded. One disadvantage of progeny testing, in a dairy context, is the long generation interval of seven years, on average, which reduces the annual rate of genetic improvement.

The four pathways of genetic improvement in dairy cattle breeding are bulls to breed bulls, BB, bulls to breed cows, BC, cows to breed bulls, CB, and cows to breed cows, CC. Each pathway uses different information to predict the breeding values of animals within the pathway.

For the BB and BC pathways, if a bull's breeding value is predicted from milk records on n daughters, then the accuracy of predicted breeding values is $\sqrt{\frac{n}{n+\lambda}}$ and $\lambda = \frac{4-h^2}{h^2}$. If the heritability of milk yield is 0.25 and there is a milk record on each of the bull's 50 daughters, then the accuracy of the bull's predicted breeding value is 0.88.

For the CB pathway, the accuracy of predicted breeding values of cows is 0.63, when the selection criterion consists of the 50 milk records of the maternal grandsire's daughters, the 50 milk records of the sire's daughters and the cow's first lactation record.

In the CC pathway, the cow may have zero, one or two lactation records, with accuracies of predicted genetic merit of 0.49, 0.63 and 0.69, respectively, giving an average accuracy of 0.60. For the cow with no lactation records, the accuracy of the predicted breeding value is

$$r_{A\hat{A}}^2 = \frac{1}{4}\left(r_S^2 + r_D^2\right)$$

where r_S^2 and r_D^2 are the accuracies of predicted genetic merit for her sire and dam. If the cow's breeding value is predicted from information on her sire's daughters and her maternal grandsire's daughters (MGS), the accuracy of the cow's predicted genetic merit is

$$\sqrt{\frac{1}{4}r_S^2 + \frac{1}{16}r_{MGS}^2} = 0.49$$

The annual rate of genetic improvement in a conventional dairy breeding scheme, using progeny testing, can be determined when information on the accuracies of predicted breeding values are combined with the selection differentials and generation intervals of each pathway, as shown in Table 9.1.

Table 9.1. Response to selection in each of the four pathways

Pathway	Average generation interval (years)	Accuracy of predicted breeding value	Proportion selected	Selection differential (σ_P)	Response (σ_P)
BB	6.5	0.88	0.03	2.27	1.00
BC	7.5	0.88	0.11	1.70	0.75
CB	6.5	0.63	0.02	2.38	0.75
CC	4.5	0.60	0.90	0.20	0.12

The response per generation in each pathway is $ir_{A\hat{A}}h\sigma_P$. The annual genetic improvement in a conventional dairy breeding scheme is the sum of the genetic improvements in each pathway divided by the sum of the generation intervals, which is $0.10\sigma_P$ (Nicholas and Smith, 1983).

In a conventional dairy breeding scheme, the two main constraints on the annual rate of genetic improvement are the long generation intervals of the bulls and cows to breed bulls, BB and CB, and the bulls to breed cows, BC, pathways, and the very low selection differential in the cows to breed cows, CC, pathway. The generation interval in the BB and BC pathways could be substantially reduced if bulls' breeding values were predicted from information on their parents and relatives, rather than from progeny test information. Secondly, if the reproductive rate of cows could be increased, then the selection differential in the cows to breed cows, CC, pathway would be increased. The combination of the two factors— (1) selecting bulls on predicted breeding value, rather than from progeny test records and (2) increasing the reproductive rate of cows— is the basis of multiple ovulation and embryo transfer (MOET) in a dairy cattle breeding scheme (Nicholas and Smith, 1983).

Multiple ovulation and embryo transfer (MOET)

Multiple ovulation to increase the number of eggs and the transfer of embryos to recipient cows increases the reproductive rate of cows and so increases the potential selection differential of cows to breed cows. Cows would be selected on a within-family basis, as a cow's transferred embryos could be considered as full-sibs, to maintain the number of cow families and prevent an increase in the rate of inbreeding.

Evaluation of bulls on the lactation records of their half-sibs and their full-sibs, through the use of MOET, will substantially reduce the generation interval for bulls, compared to progeny testing. The use of half-sib and full-sib information, rather than progeny information, will result in a lower accuracy for a bull's predicted breeding value, due to the smaller number of half-sibs and full-sibs compared to the number of progeny in a conventional breeding programme. Therefore, two alternative procedures have been proposed for dairy cattle breeding schemes using MOET. In the adult scheme, each cow has a lactation to increase the accuracy of her predicted breeding value, but the generation interval is longer than in the juvenile scheme, in which animals are selected only on the basis of parental breeding values.

Assume that each bull is mated to s cows, each of which has n male and n female progeny from transferred embryos. For a particular animal, there will be lactation records available from its dam, its dam's (n–1) full-sibs, FS, and (s–1)n half-sibs, HS, and its grandam, and similarly on the sire side of the pedigree:

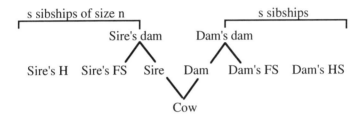

Juvenile MOET scheme

At 13 months of age, animals are selected on the basis of their parental breeding values, which incorporate lactation records on the animal's dam, the dam's full-sibs and half-sibs, and also the sire's full-sibs and half-sibs. At 15 months of age, the selected animals are mated, with a generation interval of 22 months (Nicholas and Smith, 1983). An animal and its full-sibs have the same predicted breeding values, as there is no information to differentiate between the full-sibs. The proportion of males selected is 1/s, rather than 1/sn, as full-sib males have the same predicted breeding value; so bulls are effectively selected on a between full-sib family basis. In contrast, the proportion of selected females is 1/n, as cows are selected at random, on a within full-sib basis, to replace their dams.

Example

If each bull is mated to eight cows, each of which has four male and four female progeny, through MOET, then the predicted breeding values have an accuracy of 0.43. The response to selection in the male and female pathways is as follows:

	Generation interval (months)	Accuracy of predicted breeding value	Proportion selected	Selection differential (σ_P)	Response (σ_P)
Males	22	0.43	1/8	1.65	0.36
Females	22	0.43	1/4	1.27	0.27

The annual response of a juvenile MOET scheme is $0.17\,\sigma_P$, equal to the sum of the responses in the two pathways divided by the sum of the generation intervals (Nicholas and Smith, 1983). The annual rate of genetic improvement is substantially greater than that of a conventional progeny testing scheme, but there may have to be an incentive for dairy producers to use semen from bulls that have not been progeny tested.

Adult MOET scheme

In the adult MOET scheme, cows are selected after one lactation, which increases the generation interval to 44 months, but the accuracy of predicted breeding values is also increased, as lactation records are now available on the cow's full-sibs and half-sibs.

Selection of males in MOET schemes

In both the juvenile and adult MOET schemes, there is no performance information on males, to enable differentiation between full-sib males, compared to selecting a bull at random from within a full-sib group. If full-sib bulls could be differentiated on the basis of an indicator trait, then the selection intensity of bulls would be increased (Woolliams and Smith, 1988). Incorporation of between full-sib and within full-sib selection for males changes the annual rate of response in a juvenile MOET scheme from

$$\frac{i_m r_{A\hat{A}} + i_f r_{A\hat{A}}}{L_m + L_f} h\sigma_P$$

to

$$\frac{\left(i_m r_{A\hat{A}}\right)_{between} + \left(i_m r_{A\hat{A}}\right)_{within} + \left(i_f r_{A\hat{A}}\right)}{L_m + L_f} h\sigma_P$$

where the m and f subscripts denote the selection intensities and generation intervals of males and females.

The standardised selection differential for within full-sib family selection is $i\sqrt{\frac{n}{n-1}}$ (used by Woolliams and Smith, 1988), where n is the number of full-sibs, and the accuracy of a predicted breeding value given within full-sib selection on the basis of an indicator trait, X, is

$$\tfrac{1}{2} r_A h_X \sqrt{\frac{n-1}{n} \bigg/ \left(1 - \tfrac{1}{2} h_X^2\right)}$$

The term $\left(i_m r_{A\hat{A}}\right)_{within}$ equals $\tfrac{1}{2} i_m r_A h_X \bigg/ \sqrt{1 - \tfrac{1}{2} h_X^2}$.

The above equations are based on one generation of selection and have not taken account of the reduction in genetic variance due to selection. The long term responses to MOET schemes will be lower than implied by the above equations.

Example

Blood urea nitrogen (BUN) is negatively correlated with protein deposition, with protein deposited either as muscle or as milk. Bulls could be measured for BUN to differentiate between full-sibs. As with the previous example on juvenile MOET schemes, it is assumed that each bull is mated to eight cows, each of which has four male and four female progeny, through MOET.

The annual responses in milk yield in a juvenile MOET scheme, with BUN measured on bulls, for different values of the heritability for BUN and the genetic correlation of BUN with milk yield (Woolliams and Smith, 1988) are as follows:

	Genetic correlation between BUN and milk yield	
h^2_{BUN}	0.25	0.50
0.25	$0.18\,\sigma_P$	$0.20\,\sigma_P$
0.50	$0.19\,\sigma_P$	$0.21\,\sigma_P$

For one round of selection, inclusion of an indicator trait can further increase the annual response of a juvenile MOET scheme, from $0.17\,\sigma_P$ to $0.20\,\sigma_P$, provided that the genetic correlation between the indicator trait and milk yield is at least 0.5, and the indicator trait has a heritability of 0.25.

Chapter ten

Simultaneous Prediction of Breeding Values for Several Animals

In Chapter 5, the breeding value of one animal was predicted, when information was available on relatives. Rather than predicting breeding values separately for each animal, it would be more efficient if breeding values were predicted for all animals, simultaneously. The equations derived from selection index theory to predict the breeding value of one animal, as discussed in Chapters 4, 5 and 6, are used to predict the breeding values of all animals under consideration.

For example, one animal's breeding value is predicted as

$$\hat{A} = b_{AP}(P - P_{Pop})$$

where b_{AP} is the regression coefficient of breeding value on phenotype, P is the animal's phenotype and P_{Pop} is the population mean phenotype. If b_{AP} and P were replaced by a matrix containing the regression coefficients and a vector with the animals' phenotypes, then breeding values for all animals could be predicted simultaneously. The selection objective and selection criterion would consist of all animals' breeding values and all phenotypic measurements, respectively.

Breeding value prediction

Let X be the vector of phenotypes for n animals and let μ be the population mean phenotype. Then the vector of predicted breeding values, $\hat{A}$ is

$$\hat{A} = b'(X - \mu 1)$$

where 1 is a vector of ones. With observations on the n animals, the phenotypic and genetic variance–covariance matrices, P and G, are of size n × n, with $b = P^{-1}G$. The parameter b is now a matrix of size n × n.

In terms of a selection objective and a selection criterion, the n animals with measurements are equivalent to the n traits to be improved. The variances, prediction error variances and accuracies of all of the predicted breeding values are determined in a comparable manner to when the breeding value was predicted for one animal.

Variance of predicted breeding values

The variance of a predicted breeding value for one animal, discussed in Chapter 4, is $b_{A\bar{P}}^2 \, \text{var}(\bar{P})$, where $b_{A\bar{P}}$ is the regression coefficient for additive genetic merit on the mean phenotypic measurement of the individual, $\bar{P}$. In matrix notation, the variance of a predicted breeding value is b' Pb, where P is the variance of the mean phenotypic measurement, with $b = P^{-1}G$

The variance–covariance matrix of predicted breeding values for n animals is

$$b' Pb = (P^{-1}G)' P(P^{-1}G) = (G' P^{-1})P(P^{-1}G) = G' P^{-1}G$$

The diagonal terms of $G' P^{-1}G$ are the variances of the predicted breeding values.

Prediction error variance

Similarly, the prediction error variance of one animal is $\text{var}(A) - \text{var}(\hat{A})$, as outlined in Chapter 4, which can be represented in matrix notation by $C - b' Pb$, where C is the genetic variance of the mean phenotypic measurement. Therefore, by analogy, the prediction error variance can be written as $G - G' P^{-1}G$, when traits in the selection objective are equal to the traits in the selection criterion, as then C equals G.

If measurements are made on only m animals and there are n animals, the breeding values of which are to be predicted, then G is an m × n matrix of the genetic covariances between the m phenotypic measurements and the breeding values of the n animals. C is an n × n matrix of the genetic variance–covariances for breeding values of the n animals.

The prediction error variance–covariance matrix of the predicted breeding values is $C - G' P^{-1}G$, with the prediction error variances on the diagonal.

Sire Evaluation

The predicted breeding values and prediction error variances of several sires can be determined simultaneously, from measurements on their progeny, rather than separately. Information on the genetic relationships between sires can be incorporated in the breeding value prediction, such that information from one sire's progeny can contribute to the predicted breeding value of a related sire.

Use of the sire model is illustrated with an example to demonstrate the link with selection index methodology, firstly assuming the sires are unrelated and then accounting for the genetic relationship between sires.

Example

Three unrelated sires, A, B and C, have 10, 20 and 40 half-sib progeny with progeny mean values of 170, 220 and 190, respectively. The trait has a heritability of 0.25 and a phenotypic variance of 20, and the mean phenotype for the population is 180. The predicted breeding values and the prediction error variance of the three sires are required.

The selection objective can be considered as the breeding values of the three sires, and the selection criterion to consist of the progeny mean values for the three sires. The phenotypic variance–covariance matrix for the mean measurements of the three progeny groups is:

$$P = \begin{bmatrix} \left(t+\dfrac{1-t}{10}\right)\sigma_P^2 & 0 & 0 \\ 0 & \left(t+\dfrac{1-t}{20}\right)\sigma_P^2 & 0 \\ 0 & 0 & \left(t+\dfrac{1-t}{40}\right)\sigma_P^2 \end{bmatrix}$$

and the genetic covariance matrix of the three breeding values with the three progeny phenotypic means is:

$$G = \begin{bmatrix} \dfrac{1}{2}h^2\sigma_P^2 & 0 & 0 \\ 0 & \dfrac{1}{2}h^2\sigma_P^2 & 0 \\ 0 & 0 & \dfrac{1}{2}h^2\sigma_P^2 \end{bmatrix}$$

The traits in the selection objective, the breeding values of the three sires, are not equal to the traits in the selection criterion, the phenotypic mean measurements of the three groups of progeny, such that the genetic variance–covariance matrix of the three sires' breeding values, C, is equal to 2G. The matrix of regression coefficients, $b = P^{-1}G$, is

$$\begin{bmatrix} 0.80 & 0 & 0 \\ 0 & 1.14 & 0 \\ 0 & 0 & 1.45 \end{bmatrix}$$

and the predicted breeding values of sires A, B and C are

$$\hat{A} = b'(\overline{X}-\mu) = b'\begin{bmatrix} 170-180 \\ 220-180 \\ 190-180 \end{bmatrix} = \begin{bmatrix} -8.0 \\ 45.7 \\ 14.5 \end{bmatrix}$$

The prediction error variances of the three predicted breeding values are the diagonal elements of the prediction error variance–covariance matrix, $C - G'P^{-1}G$:

$$PEV = \begin{bmatrix} 3.00 \\ 2.14 \\ 1.36 \end{bmatrix}$$

The breeding values and prediction error variance are exactly the same as if they had been calculated individually rather than simultaneously. The real benefit of breeding value prediction for several animals simultaneously is when genetic relationships between individuals can be incorporated in the prediction of breeding values, which would not have been possible if breeding values were predicted separately.

If sires A and B are half-sibs, then the progeny of sire A can provide information towards the predicted breeding value of sire B and vice versa. The phenotypic mean measurements of progeny from sires A and B are no longer uncorrelated and similarly, the breeding values of sires A and B are now correlated. There is also a covariance between the breeding value of sire A with the phenotypic mean measurement of progeny from sire B and vice versa:

```
         GS
    1/2 /  \ 1/2
      S_A   S_B
    1/2|   1/2|
      X̄_A   X̄_B
```

GS: grandsire of progeny

S_A and S_B: sire A and sire B

$\bar{X}_A$ and $\bar{X}_B$: mean phenotype of progeny of sire A and sire B

The covariance between the progeny mean measurements for sires A and B is the covariance between one progeny of sire A with one progeny of sire B, which is $\frac{1}{16}h^2\sigma_P^2$. The coefficient of $\frac{1}{16}$ is $\left(\frac{1}{2}\right)^4$, as there are four steps linking progeny from sire A to progeny from sire B. The covariance between the breeding value of sire A with the phenotypic mean measurement of progeny from sire B is $\frac{1}{8}h^2\sigma_P^2$, as there are three steps connecting sire A to the progeny of sire B.

The phenotypic variance–covariance of the mean progeny phenotypes, the measured traits, and the genetic covariance matrix between the mean progeny phenotypes and the sire breeding values, the traits to be improved, are

$$P = \begin{bmatrix} \left(t+\frac{1-t}{10}\right)\sigma_P^2 & \frac{1}{16}h^2\sigma_P^2 & 0 \\ \frac{1}{16}h^2\sigma_P^2 & \left(t+\frac{1-t}{20}\right)\sigma_P^2 & 0 \\ 0 & 0 & \left(t+\frac{1-t}{40}\right)\sigma_P^2 \end{bmatrix} \text{ and } G = \begin{bmatrix} \frac{1}{2} & \frac{1}{8} & 0 \\ \frac{1}{8} & \frac{1}{2} & 0 \\ 0 & 0 & \frac{1}{2} \end{bmatrix}h^2\sigma_P^2$$

The third matrix equal to the genetic variance–covariance of the sire breeding values, C, is 2G, as the genetic relationship between progeny and sire is 0.5.

The matrix of regression coefficients, $b = P^{-1}G$, is

$$\begin{bmatrix} 0.78 & 0.09 & 0 \\ 0.17 & 1.13 & 0 \\ 0 & 0 & 1.45 \end{bmatrix}$$

and the predicted breeding values of sires A, B and C are

$$\hat{A} = b'(X-\mu) = \begin{bmatrix} -4.3 \\ 43.5 \\ 14.5 \end{bmatrix}$$

The prediction error variances of the three predicted breeding values are

$$PEV = C - G'P^{-1}G = \begin{bmatrix} 2.93 \\ 2.12 \\ 1.36 \end{bmatrix}$$

Prediction of the breeding value of sire A now incorporates the mean phenotype of progeny from sire B, such that the extra information reduces the prediction error variance. The predicted breeding value of sire A is

$$0.78(170-\mu) + 0.17(220-\mu) + 0(190-\mu)$$

The proportional reduction in the prediction error variance of 0.02 for sire A indicates that there is a contribution of information from the mean phenotype of progeny from sire B to the predicted breeding value of sire A. Given that there is information on relatives, it is appropriate to incorporate that information in the prediction of the breeding value, to increase the precision of the predicted breeding value. The predicted breeding value and prediction error variance for sire C are unchanged, as sire C is unrelated to either sire A or sire B.

The predicted breeding values of sires A and B have changed, with a larger change for sire A, when information from the progeny of each sire's half-sib was included in the prediction. The example indicates that the inclusion of additional information can change predicted breeding values, such that predicted breeding values can decrease, as for sire B, as well as increase, as with sire A.

Animal Evaluation

Rather than just predicting breeding values for sires of animals, the breeding values of all animals in a pedigree can be predicted, including those animals with no measurements. The methodology is the same as with sire evaluation, in that the phenotypic variance–covariance matrix of observations, P, the genetic covariance of an animal's phenotypes with the breeding values of all animals in the pedigree, G, and the genetic variance–covariance matrix of breeding values for all animals in the pedigree, C, are required. Animal evaluation can be considered as the selection objective, being equal to the breeding values of all animals in the pedigree, with the selection criterion consisting of the phenotypes of animals with measurements.

Example

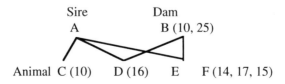

Simultaneous prediction of breeding value for several animals 119

In the pedigree of six animals, C and D are half-sibs, and D and E are full-sibs, while F is unrelated to the other animals. The measurements on each animal are given in parentheses. The heritability of the trait is 0.25, the repeatability is 0.4, the phenotypic variance is 4.0 and the population mean phenotype is 15.0. Determine the predicted breeding values for each animal.

The phenotypic variance–covariance matrix, P, of measurements is of size 4×4, as only four animals have measurements, such that

$$P = \begin{bmatrix} r_e + \frac{1-r_e}{2} & 0 & \frac{1}{2}h^2 & 0 \\ 0 & 1 & \frac{1}{4}h^2 & 0 \\ \frac{1}{2}h^2 & \frac{1}{4}h^2 & 1 & 0 \\ 0 & 0 & 0 & r_e + \frac{1-r_e}{3} \end{bmatrix} \sigma_P^2 \text{ for animals } \begin{bmatrix} B \\ C \\ D \\ F \end{bmatrix}$$

Animals B and D are dam and offspring, with a covariance of $\frac{1}{2}h^2\sigma_P^2$, while animals C and D are half-sibs and the covariance between measurements is $\frac{1}{4}h^2\sigma_P^2$. The genetic covariance matrix of four measurements with six breeding values is as follows:

$$G = \begin{bmatrix} 1 & 0 & \frac{1}{2} & 0 & 0 & \frac{1}{2} \\ 0 & 1 & \frac{1}{4} & 0 & \frac{1}{4} & \frac{1}{4} \\ \frac{1}{2} & \frac{1}{4} & 1 & 0 & \frac{1}{2} & \frac{1}{2} \\ 0 & 0 & 0 & 1 & 0 & 0 \end{bmatrix} h^2\sigma_P^2, \text{ with rows equal to animals } \begin{bmatrix} B \\ C \\ D \\ F \end{bmatrix}$$

and animals B, C, D, F, A and E in columns. Calculation of the prediction error variances also requires the genetic variance-covariance of the breeding values, C.

The matrix of regression coefficients, $b = P^{-1}G$, is

$$\begin{bmatrix} 0.34 & -0.01 & 0.14 & 0 & -0.02 & 0.16 \\ -0.01 & 0.25 & 0.05 & 0 & 0.12 & 0.06 \\ 0.08 & 0.05 & 0.23 & 0 & 0.12 & 0.10 \\ 0 & 0 & 0 & 0.42 & 0 & 0 \end{bmatrix}$$

The predicted breeding values and prediction error variances for the animals are

$$\begin{bmatrix} B \\ C \\ D \\ F \\ A \\ E \end{bmatrix} = \begin{bmatrix} 0.96 \\ -1.21 \\ 0.33 \\ 0.14 \\ -0.52 \\ 0.22 \end{bmatrix} \text{ and } \begin{bmatrix} 0.62 \\ 0.74 \\ 0.69 \\ 0.58 \\ 0.88 \\ 0.86 \end{bmatrix}$$

In the b matrix, the regression coefficients for each animal are in columns, with the underlined terms equal to the regression coefficients for the animal's own measurements. Therefore, the predicted breeding value of animal B is

$$0.34\left(\frac{20+15}{2}-15\right)-0.01(10-15)+0.08(16-15)+0\left(\frac{14+17+15}{3}-15\right)$$

Records on animals C and D have similar regression coefficients, 0.1175 and 0.1203, for predicting the breeding value of their sire, animal A, since both animals have the same number of records. When predicting the breeding value of animal E, the regression coefficients for animal E's dam B, 0.16, half-sib C, 0.06, and full-sib D, 0.10, reflect the combination of a greater number of records on the dam than on the full-sib and the genetic relationships between animal E with its relatives. The predicted breeding value of animal E is also equal to the average predicted breeding value of its parents, A and B, as expected, since animal E has no record. As animal F is unrelated to the other five animals, then its breeding value is predicted from its own records.

The prediction error variances reflect the contribution of information from animals in the pedigree to the calculation of predicted breeding values. For example, animals C and D both had one record, but information from D's dam and full-sib reduced the prediction error variance of animal D relative to animal C. Similarly, both animal A and E have no records, but there is relatively more information from animal E's relatives than from animal A's progeny, and so the prediction error variance of animal E is lower than for animal A.

The link between selection index methodology and simultaneous prediction of breeding values for several animals has been established. However, one assumption of selection index methodology is that all animals are measured in the same environment. Obviously, such an assumption is not realistic, such that different environmental effects need to be accounted for when predicting breeding values, which is the subject of the next chapter.

Chapter eleven

Prediction of Breeding Values and Environmental Effects

In previous chapters, animals were assumed to belong to the same environmental group, and phenotypic differences between animals were due to genetic differences, rather than to a combination of genetic and environmental differences. Examples of environmental effects are year of measurement, management system, age of dam or the age at the end of performance test. One approach would be to estimate the environmental effects and then adjust each animal's record for the estimated environmental effects, before predicting breeding values.

Table 11.1. Average milk yields of daughters by sire and herd

	Sire		
Farm	1	2	3
A	27 (24)	23 (20)	
B		17 (20)	16 (19)

For example, if sires have the same number of daughters on each farm, then between-farm differences can be accounted for in the comparison of sires. If each of three sires have a very large number of daughters on two farms and the average daily milk yields are as shown in Table 11.1, then the difference of 6 kg/day between progeny of sire 2 on farms A and B provides an estimate of the effect of being reared on farm A or B. Therefore, the value of 3 kg/day is subtracted from progeny means in farm A, but added to progeny means in farm B, such that the "population" mean after accounting for farms is $(24 + 20 + 19)/3 = 21$ kg/day. After adjusting the progeny mean values for the effects of farm, as indicated by values in parentheses, the breeding values of the sires can be predicted as 6, −2 and −4 kg/day, respectively. If there was a high degree of confounding between genotype and environment, then such an approach would be unsatisfactory. If the number of progeny of each sire on each farm is not very large, then it would be preferable to predict sires' breeding values and estimate the environmental effects simultaneously.

A procedure for simultaneous prediction of breeding values and estimation of environmental effects, called best linear unbiased prediction, was developed by Henderson (1953).

Best Linear Unbiased Prediction (BLUP)

BLUP can be used with different models to predict breeding values and estimate environmental effects. The properties of the BLUP procedure are as follows:
Best: maximisation of the correlation between the true breeding value and the predicted breeding value
Linear: predicted breeding values are linear functions of the observations
Unbiased: estimates of fixed effects are unbiased and the unknown, true breeding values are distributed about the predicted breeding values
Prediction: the procedure predicts the true breeding values

BLUP is generally used to predict sire breeding values, given measurements on progeny, or to predict breeding values of animals with repeated records, or to predict breeding values of all animals in the pedigree. The three models are called the sire model, the repeatability model and the individual animal model, respectively. Use of BLUP in each of the three models will be discussed in turn.

Fixed and random effects

The environmental effects, such as farm of measurement, sex or breed of animal and age of dam, are referred to as the fixed effects, as the effect is of specific interest. Hypotheses can be proposed regarding the fixed effects, such as "There is no difference in mean perforamance of animals from the two breeds A and B". A model with a fixed effect is represented as:

$$y_{ij} = \mu + \alpha_i + e_{ij}$$

which refers to the jth animal in the ith class of the fixed effect. A model which estimates only fixed effects is called a fixed effects model.

In contrast, estimating the effect of the specific pen in which an animal was housed during performance test is not of interest, although some account should be taken of variation in the performance test traits, due to animals being tested in different pens. Effects, such as pen, are referred to as random effects. A model with a random effect would be :

$$y_{ij} = \mu + \delta_i + e_{ij}$$

which refers to the jth animal in the ith class of the random effect. A random effects model estimates variation between the random effects. A mixed model includes both fixed and random effects, other than the error term.

In both fixed and random effects models, the error terms are random effects with a mean of zero and a variance of σ_e^2. In the model with a fixed effect, the α's represent differences between classes of the fixed effect, while in the model with a random effect, the δ's are random samples from a population with a mean of zero and a variance of σ_δ^2. The covariance between the random effect and the error term is assumed to be zero. The subject of fixed and random effects is discussed by Searle (1971).

In animal breeding, the sires or animals, the breeding values of which are to be predicted, are referred to as random effects, as the sample of sires in the data are assumed to belong to a population of sires with a variance of σ_s^2. However, the BLUP method is used to estimate the sire effects, or predict the sires' breeding values, such that the sires are also "treated" as fixed effects.

Sire Model

In the sire model, sire effects equal to half the breeding values of sires are predicted from observations on progeny. The sire model is particularly used in dairy cattle breeding programmes, where progeny testing of sires is practised. The model for the observation on one animal, in this case the offspring of a sire, is

$$Y = A + E$$

where A is half the breeding value of the sire and E is the environmental effect.

The equation for the sire model is

$$y = Xb + Zu + e$$

where y is the vector of observations, X is the incidence matrix of the fixed effects, b is the vector of fixed effects, Z is the incidence matrix of the sire effects, u is the vector of sire effects and e is the vector of residuals. In an incidence matrix, each element consists of either a zero or a one, according to which level of the fixed or random effect each animal is classified.

If all effects are considered as fixed, the following equations are derived:

X'Xb + X'Zu = X'y   from multiplying both sides of the equation by X'
Z'Xb + Z'Zu = Z'y   from multiplying both sides of the equation by Z'

or

$$\begin{bmatrix} X'X & X'Z \\ Z'X & Z'Z \end{bmatrix} \begin{bmatrix} \hat{b} \\ \hat{u} \end{bmatrix} = \begin{bmatrix} X'y \\ Z'y \end{bmatrix}$$

The mixed model equations should include a factor relating to the variation between the sire effects, as sires are random effects and are not fixed, and this can be derived from selection index methodology.

The predicted breeding value of a sire, $\hat{A}$, given the mean progeny measurement, $\overline{P}$, and the population mean, μ, is

$$\hat{A} = b(\overline{P} - \mu) \quad \text{where} \quad b = P^{-1}G$$

or

$$\left[\sigma_s^2 + \frac{1}{n}\sigma_e^2\right]\hat{A} = \left[\frac{1}{4}h^2 + \frac{1}{n}\left(1 - \frac{1}{4}h^2\right)\right]\sigma_P^2 \hat{A} = \frac{1}{2}h^2 \sigma_P^2 \overline{P}$$

assuming that $\mu = 0$, and by mutiplying throughout with $\dfrac{n}{\sigma_s^2} = \dfrac{4n}{h^2 \sigma_P^2}$, gives

$$[n + \lambda]\hat{u} = n\overline{P} = \left[\sum_{i=1}^{n} P_i\right]$$

where $\lambda = \dfrac{4-h^2}{h^2}$ and $\hat{u}$ is half the predicted genetic merit of the sire. However, if the estimated sire effect is determined from the sire part of the above fixed effects model

$$[Z'Z][\hat{u}] = [Z'y] \quad \text{or} \quad [n][\hat{u}] = \left[\sum_{i=1}^{n} P_i\right]$$

then the solution for the sire effect would not be the same as from selection index theory, as the term λ is omitted from the left-hand side of the equation.

The equation from selection index theory is equivalent to incorporating the λI matrix in the fixed effects model, so that the mixed model equations include a factor relating to the variation between the sires (Henderson, 1973):

$$\begin{bmatrix} X'X & X'Z \\ Z'X & Z'Z + \lambda I \end{bmatrix} \begin{bmatrix} \hat{b} \\ \hat{u} \end{bmatrix} = \begin{bmatrix} X'y \\ Z'y \end{bmatrix}$$

In Chapter 10, genetic relationships between animals were included in the P and G matrices, such that information from relatives could be combined for prediction of the individual's breeding value. The matrix of genetic relationships between sires, A, can be incorporated into the equation as:

$$\left[A\sigma_s^2 + \frac{1}{n} I \sigma_e^2\right] \hat{A} = A \frac{1}{2} h^2 \sigma_P^2 \overline{P}$$

and multiplying thoughout with $\dfrac{n}{\sigma_s^2} A^{-1}$, gives

$$\left[n + \lambda A^{-1}\right] \hat{u} = n\overline{P} = \left[\sum_{i=1}^{n} P_i\right]$$

In analogous manner, relationships between sires are also incorporated into the mixed model equations by including the term λA^{-1} to obtain the mixed model equations:

$$\begin{bmatrix} X'X & X'Z \\ Z'X & Z'Z + \lambda A^{-1} \end{bmatrix} \begin{bmatrix} \hat{b} \\ \hat{u} \end{bmatrix} = \begin{bmatrix} X'y \\ Z'y \end{bmatrix}$$

where A is the numerator relationship matrix consisting of the genetic relationships between sires. In the context of a selection objective and selection criterion, the A matrix is analogous to the genetic variance–covariance matrix of sire breeding values divided by the genetic variance, $C\sigma_A^{-2}$.

Computational procedures to solve the mixed model equations for simultaneous estimation of the fixed or environmental effects and prediction of breeding values are outwith the remit of this text, but are comprehensively discussed by Mrode (1996).

Prediction of breeding values and environmental effects

Example

The milk yields of 13 Jersey cows from two sires were recorded on two farms for one day. The heritability of daily milk yield is 0.25. The breeding values of the two sires are to be predicted. The animal identities, farm, sire and each animal's milk yield are as follows:

Animal	Farm	Sire	Milk yield
1	A	1	8
2	A	1	9
3	A	2	11
4	A	2	12
5	A	2	12
6	A	2	13
7	A	2	14
8	B	1	15
9	B	1	14
10	B	1	15
11	B	2	18
12	B	2	19
13	B	2	20

The mixed model equations require the incidence matrices for the farm, X, and sire, Z, effects, which are obtained from the data and each element consists of either a zero or a one, according to which farm or sire each animal was classified. Similarly, the vector of observations, y, is required.

Animal	X matrix Farm A	X matrix Farm B	Z matrix Sire 1	Z matrix Sire 2	y vector Milk
1	1	0	1	0	8
2	1	0	1	0	9
3	1	0	0	1	11
4	1	0	0	1	12
5	1	0	0	1	12
6	1	0	0	1	13
7	1	0	0	1	14
8	0	1	1	0	15
9	0	1	1	0	14
10	0	1	1	0	15
11	0	1	0	1	18
12	0	1	0	1	19
13	0	1	0	1	20

The mean daily milk yield and number of observations, in parentheses, for each farm–sire subclass are as follows:

	Sire 1	Sire 2	Total
Farm A	8.5 (2)	12.4 (5)	79 (7)
Farm B	14.6 (3)	19.0 (3)	101 (6)
Total	61 (5)	119 (8)	

The left-hand side of the mixed model equations can be split into the submatrices that describe the number of observations in each class of the fixed effects, X'X, the random effects, Z'Z, and the cross-classification of fixed and random effects, X'Z,

$$X'X = \begin{bmatrix} 7 & 0 \\ 0 & 6 \end{bmatrix}, \quad Z'Z = \begin{bmatrix} 5 & 0 \\ 0 & 8 \end{bmatrix}, \quad X'Z = \begin{bmatrix} 2 & 5 \\ 3 & 3 \end{bmatrix}$$

Note that the submatrices can be simply derived from the number of observations for each farm–sire subclass and that, unlike selection index methodology, information on the population mean is not required for breeding value prediction.

The right-hand side of the mixed model equation is the sum of the observations in each class of the fixed effects, X'y, and the random effects, Z'y,

$$X'y = \begin{bmatrix} 79 \\ 101 \end{bmatrix}, \quad Z'y = \begin{bmatrix} 61 \\ 119 \end{bmatrix}$$

In the sire model, the observations on each animal are not used directly, as only the sum of observations in each subclass of fixed and random effects is included in the mixed model equations. Later in this chapter, discussion of the individual animal model will illustrate how the observations on each animal can be directly included in the mixed model equations.

With sires treated as fixed effects, the fixed model equations are

$$\begin{bmatrix} 7 & 0 & 2 & 5 \\ 0 & 6 & 3 & 3 \\ 2 & 3 & 5 & 0 \\ 5 & 3 & 0 & 8 \end{bmatrix} \begin{bmatrix} \hat{b} \\ \hat{u} \end{bmatrix} = \begin{bmatrix} 79 \\ 101 \\ 61 \\ 119 \end{bmatrix} \quad \text{with solutions} \quad \begin{bmatrix} \hat{b} \\ \hat{u} \end{bmatrix} = \begin{matrix} \text{farm A} \\ \text{farm B} \\ \text{sire 1} \\ \text{sire 2} \end{matrix} = \begin{bmatrix} 10.40 \\ 16.83 \\ 11.56 \\ 15.68 \end{bmatrix}$$

With sires treated as random effects, the inverse of the numerator relationship matrix needs to be incorporated into the mixed model equations. If the two sires are unrelated, then $A = \begin{bmatrix} 1 & 0 \\ 0 & 1 \end{bmatrix}$.

If the heritability is 0.25, then $\lambda = \dfrac{\sigma_e^2}{\sigma_s^2} = \dfrac{1 - \frac{1}{4}h^2}{\frac{1}{4}h^2} = \dfrac{4 - h^2}{h^2} = 15$

and $\lambda A^{-1} = \begin{bmatrix} 15 & 0 \\ 0 & 15 \end{bmatrix}$

The mixed model equations are

$$\begin{bmatrix} 7 & 0 & 2 & 5 \\ 0 & 6 & 3 & 3 \\ 2 & 3 & 20 & 0 \\ 5 & 3 & 0 & 23 \end{bmatrix} \begin{bmatrix} \hat{b} \\ \hat{u} \end{bmatrix} = \begin{bmatrix} 79 \\ 101 \\ 61 \\ 119 \end{bmatrix} \text{ with solutions } \begin{bmatrix} \hat{b} \\ \hat{u} \end{bmatrix} = \begin{bmatrix} \text{farm A} \\ \text{farm B} \\ \text{sire 1} \\ \text{sire 2} \end{bmatrix} = \begin{bmatrix} 11.04 \\ 16.83 \\ -0.58 \\ 0.58 \end{bmatrix}$$

Although the solutions for the farms are similar to the fixed effects model, the sire solutions now sum to zero.

In the mixed model equations, no account was taken of the genetic relationship between the sires and their offspring, on whom the milk yields were recorded. The sire solutions are equal to one half of the sires' predicted breeding values, as the genetic relationship between sire and offspring is one half.

Information on the genetic relationship between sires can be incorporated into the mixed model equations, in a manner similar to that outlined in Chapter 10.

If sire 1 is a parent of sire 2, the numerator relationship matrix is

$$\begin{bmatrix} 1 & 0.5 \\ 0.5 & 1 \end{bmatrix} \text{ such that } \lambda A^{-1} = \begin{bmatrix} 20 & -10 \\ -10 & 20 \end{bmatrix} \text{ and } Z'Z + \lambda A^{-1} = \begin{bmatrix} 25 & -10 \\ -10 & 28 \end{bmatrix}$$

Solutions to the mixed model equations are now obtained from

$$\begin{bmatrix} 7 & 0 & 2 & 5 \\ 0 & 6 & 3 & 3 \\ 2 & 3 & 25 & -10 \\ 5 & 3 & -10 & 28 \end{bmatrix} \begin{bmatrix} \hat{b} \\ \hat{u} \end{bmatrix} = \begin{bmatrix} 79 \\ 101 \\ 61 \\ 119 \end{bmatrix} \text{ and } \begin{bmatrix} \hat{b} \\ \hat{u} \end{bmatrix} = \begin{bmatrix} \text{farm A} \\ \text{farm B} \\ \text{sire 1} \\ \text{sire 2} \end{bmatrix} = \begin{bmatrix} 11.14 \\ 16.83 \\ -0.34 \\ 0.34 \end{bmatrix}$$

Note that after accounting for the genetic relationships between sires, that the sire solutions still sum to zero.

Incorporation of information on the genetic relationships between sires, for breeding value prediction, is based on the assumption that the sires are a sample from a population with mean breeding value of zero. However, there are occasions when groups of animals have been derived from populations with different mean breeding values. For example, in dairy cattle, the progeny of Holstein cattle from North America may have a different mean breeding value than progeny of Holstein cattle from Great Britain. Therefore, the existence of different genetic groups should be incorporated in the methodology for breeding value prediction. Mrode (1996) discusses the procedure for grouping, based on the methods of Thompson (1979) and Westell et al. (1988).

Several computer packages are available for BLUP evaluations, such as PEST (Groeneveld et al., 1990).

Residual Maximum Likelihood (REML)

Predicted breeding values determined from solving the mixed model equations are unbiased, provided that the genetic and phenotypic parameters, and therefore the value of λ, are known without error. There are several methods to estimate the required parameters, particularly for unbalanced experimental designs, as is generally the case with animal breeding data. ANOVA procedures are based on the assumption that animals are a random sample from the population, which will not be approriate in selected populations, such as in livestock improvement programmes. Maximum likelihood methods can account for selection conditional on all the information, which has contributed to selection decisions, being included in the analysis. Even when the condition is not fully satisfied, parameters estimated with maxiumum likelihood are generally less biased than ANOVA estimates (Meyer and Thompson, 1984).

One maximum likelihood procedure is REML (Patterson and Thompson, 1971), which is widely used for estimation of genetic and phenotypic parameters. REML is an iterative procedure, involving two steps:

(1) For a univariate analysis, the value of λ is calculated from estimates of the genetic and phenotypic variances and the mixed model equations are solved for the fixed effects and predicted breeding values, $\hat{u}$.

(2) Updated estimates of the genetic and phenotypic variances are determined from the predicted breeding values.

The cycle is repeated until estimates of the genetic and phenotypic parameters converge.

To illustrate the relationship between predicted breeding values, $\hat{u}$, of s sires, and estimates of the genetic and phenotypic variances, one such relationship is:

$$\hat{\sigma}_s^2 = \left[\hat{u}' A^{-1}\hat{u} + \sigma_e^2 \text{tr}(A^{-1}C)\right]/s$$

and

$$\hat{\sigma}_e^2 = \left[y' y - y' X\hat{b} - y' Z\hat{u}\right]/[N - r(X)]$$

where $C = \left[Z' MZ + \lambda A^{-1}\right]^{-1}$, $M = I - X'(X' X)^{-1}X$, $\text{tr}\left[A^{-1}C\right]$ is the sum of the diagonal elements of the matrix $A^{-1}C$, N is the number of observations and r(X) is the column rank of the X matrix (Harville, 1977).

Derivation of the expected value of the product of predicted breeding values is illustrated for s unrelated sires, each with n half-sib progeny, and no fixed effects. The mixed model equations reduce to a random effects model:

$$\left[Z' Z + \lambda I^{-1}\right][\hat{u}] = [Z' y]$$

For one sire

$$\hat{u}_i = \left[Z_i' Z_i + \lambda I_i^{-1}\right]^{-1} Z_i' y_i = C_i Z_i' y_i$$
$$= [n + \lambda]^{-1} n\bar{P}$$
$$= b\bar{P}$$

where b is the regression coefficient of the estimated sire effect on the progeny mean phenotype.

The variance of the estimated sire effects is

$$\frac{1}{s}\hat{u}' \hat{u} = b^2 \text{var}(\bar{P})$$

However,

$$\text{var}(\overline{P}) = \left[t + \frac{1-t}{n}\right]\sigma_P^2 = \left[\sigma_s^2 + \frac{\sigma_e^2}{n}\right] = \sigma_s^2\left[\frac{n+\lambda}{n}\right] = \frac{\sigma_s^2}{b} = \sigma_s^2\left[\frac{n+\lambda}{n}\right]$$

such that

$$\frac{1}{s}\hat{u}'\hat{u} = \sigma_s^2\left[1 - \frac{\lambda}{n+\lambda}\right]$$

and

$$\sigma_s^2 = \frac{1}{s}\left[\hat{u}'\hat{u} + \sigma_e^2[n+\lambda]^{-1}\right]$$

which has the same structure as the formula for the REML estimate of σ_s^2

$$\hat{\sigma}_s^2 = \frac{1}{s}\left[\hat{u}'A^{-1}\hat{u} + \sigma_e^2\text{tr}(A^{-1}C)\right]$$

Computer packages for REML analysis with one random effect or for several random effects using the DFREML (Graser et al., 1987) procedure are available (Meyer, 1985, 1989).

Repeatability Model

Breeding values of animals with repeated measurements and specific environmental effects can be predicted in an manner analogous to predicting sire breeding values with measurements on progeny. The model with one observation is

$$Y = A + E_S + E$$

where A is the breeding value of the animal, E_S is the specific environmental effect, as discussed in Chapter 3, and E is the general environmental effect, which will affect all animals. The mixed model equations are essentially formed in the same manner as with the sire model, but the definition of λ changes to

$$\lambda = \frac{\sigma_E^2}{\sigma_A^2} = \frac{1-r_e}{h^2}$$

The value of the repeatability, r_e, is incorporated in the model by including the term $Z'Z + \gamma I$ in the mixed model equations, where

$$\gamma = \frac{\sigma_E^2}{\sigma_{E_S}^2} = \frac{\sigma_E^2}{(r_e - h^2)\sigma_P^2} = \frac{1-r_e}{r_e - h^2}$$

and I is the identity matrix, with dimensions equal to the number of animals. The inclusion of $Z'Z + \gamma I$ in the mixed model equations requires an extra row and an extra column in the mixed model equations, which are

$$\begin{bmatrix} X'X & X'Z & X'Z \\ Z'X & Z'Z + \lambda A^{-1} & Z'Z \\ Z'X & Z'Z & Z'Z + \gamma I \end{bmatrix} \begin{bmatrix} \hat{b} \\ \hat{u}_A \\ \hat{u}_r \end{bmatrix} = \begin{bmatrix} X'y \\ Z'y \\ Z'y \end{bmatrix}$$

where u_A and u_r are the additive genetic effect and the specific animal effect, respectively. Prediction of the animal's future performance requires u_A and u_r, while prediction of the animal's offspring performance is based only on u_A.

Example

The data set used to illustrate the sire model is adapted for the repeated measurements model. The fixed effect of farm is replaced by month of measurement and the random effect of sire is replaced by the animal effect. With the animals assumed to be full-sibs and a repeatability of 0.4, the sub-matrices λA^{-1} and γI are $\begin{bmatrix} 3.2 & -1.6 \\ -1.6 & 3.2 \end{bmatrix}$ and $\begin{bmatrix} 4 & 0 \\ 0 & 4 \end{bmatrix}$, respectively, such that the mixed model equations are

$$\begin{bmatrix} 7 & 0 & 2 & 5 & 2 & 5 \\ 0 & 6 & 3 & 3 & 3 & 3 \\ 2 & 3 & 8.2 & -1.6 & 5 & 0 \\ 5 & 3 & -1.6 & 11.2 & 0 & 8 \\ 2 & 3 & 5 & 0 & 9 & 0 \\ 5 & 3 & 0 & 8 & 0 & 12 \end{bmatrix} \begin{bmatrix} \hat{b} \\ \hat{u}_a \\ \hat{u}_r \end{bmatrix} = \begin{bmatrix} 79 \\ 101 \\ 61 \\ 119 \\ 61 \\ 119 \end{bmatrix} \text{ and solutions } \begin{bmatrix} \text{month 1} \\ \text{month 2} \\ \text{animal 1} \\ \text{animal 2} \\ \text{repeat 1} \\ \text{repeat 2} \end{bmatrix} = \begin{bmatrix} 10.64 \\ 16.83 \\ -0.68 \\ 0.68 \\ -0.82 \\ 0.82 \end{bmatrix}$$

The predicted breeding values of the two animals are −0.68 and 0.68, respectively, but the predicted future performance of the animals is −1.50 and 1.50. Note that the predicted breeding values of the two animals are double the sire solutions estimated in the sire model, in the previous section. The sire model predicted half of each sire's genetic merit, based on progeny measurements, which is analogous to repeated measurements on the sire itself, except that the repeated measurements model predicts the genetic merit of each animal.

Individual Animal Model

The individual animal model or the animal model is the most general model, as repeated measurements can be incorporated and breeding values can be predicted for all animals in the pedigree, even those animals without records. Construction of the mixed model equations is the same as for the repeated measurements model, except that the numerator relationship matrix, A, contains the genetic relationships between all animals in the pedigree and the X'Z and Z'Z matrices are extended with rows and columns of zeros to accommodate animals without records. Solving the mixed model equations for the individual animal model, where the dimensions of the Z matrix equal the number of animals in the pedigree, requires considerable computing power. Improvement in the efficiency of computational procedures for solving the mixed model equations is the focus of several research projects, particularly with ever increasing numbers of equations as models become more complex, with multivariate analyses, inclusion of common environmental and maternal effects and particular traits being recorded on a subset of animals.

Example

Milk yield was measured on three cows during three months. The heritability, repeatability and phenotypic variance are 0.4, 0.6 and 10, respectively. Predict the breeding values of the cows, their sire and dam, and determine the prediction error variance of the predicted breeding values and of the cows' future performance. The pedigree of the animals and the milk records, by month of measurement are:

Month	Cow 1	Cow 2	Cow 3
1	10		9
2	12		10
3		15	12

The submatrix, corresponding to the fixed effect of month, X'X, is

$$X'X = \begin{bmatrix} 2 & 0 & 0 \\ 0 & 2 & 0 \\ 0 & 0 & 2 \end{bmatrix}$$

as there are two observations in each month.

The submatrix cross-classifying the fixed and random effects, X'Z, is a 3×5 matrix, as there are three months and five animals in the pedigree. The columns of X'Z corresponding to animals without records consist of zeros:

$$X'Z = \begin{bmatrix} 1 & 0 & 1 & 0 & 0 \\ 1 & 0 & 1 & 0 & 0 \\ 0 & 1 & 1 & 0 & 0 \end{bmatrix}$$

The submatrix for the random effects, Z'Z, is a 5×5 matrix, as there are five animals in the pedigree, and the rows and columns corresponding to animals without records consist of zeros:

$$Z'Z = \begin{bmatrix} 2 & 0 & 0 & 0 & 0 \\ 0 & 1 & 0 & 0 & 0 \\ 0 & 0 & 3 & 0 & 0 \\ 0 & 0 & 0 & 0 & 0 \\ 0 & 0 & 0 & 0 & 0 \end{bmatrix}$$

The upper triangle of the symmetric numerator relationship matrix A is:

$$A = \begin{bmatrix} 1 & \frac{1}{2} & \frac{1}{4} & \frac{1}{2} & \frac{1}{2} \\ & 1 & \frac{1}{4} & \frac{1}{2} & \frac{1}{2} \\ & & 1 & 0 & \frac{1}{2} \\ & & & 1 & 0 \\ & & & & 1 \end{bmatrix}$$

The inverse of the A matrix can be derived explicitly from the pedigree, rather than having to invert the A matrix (Henderson, 1976). For each animal in the pedigree, the elements of A^{-1} are added in sequence, using the rules:

Animal	Sire	Dam	Add to element (and its transpose) in A^{-1}					
(a)	(s)	(d)	(a,a)	(a,s)	(a,d)	(s,s)	(d,d)	(s,d)
	Known	Known	2	−1	−1	1/2	1/2	1/2
	Known	—	4/3	−2/3		1/3		
	—	Known	4/3		−2/3		1/3	
	—	—	1					

Note that a procedure to account for inbreeding, when calculating the inverse of the A matrix, has been developed by Quaas (1976).

The A^{-1} matrix for the pedigree in the example is

$$A^{-1} = \begin{bmatrix} 2 & 0 & 0 & -1 & -1 \\ 0 & 2 & 0 & -1 & -1 \\ 0 & 0 & 4/3 & 0 & -2/3 \\ -1 & -1 & 0 & 1/2+1/2+1 & 1/2+1/2 \\ -1 & -1 & -2/3 & 1/2+1/2 & 1/2+1/2+1/3+1 \end{bmatrix}$$

The mixed model equations are

$$\begin{bmatrix} X'X & X'Z & X'Z \\ Z'X & Z'Z+\lambda A^{-1} & Z'Z \\ Z'X & Z'Z & Z'Z+\gamma I \end{bmatrix} \begin{bmatrix} \hat{b} \\ \hat{u}_A \\ \hat{u}_r \end{bmatrix} = \begin{bmatrix} X'y \\ Z'y \\ Z'y \end{bmatrix}$$

with $\lambda = \dfrac{1-r_e}{h^2} = 1$ and $\gamma = \dfrac{1-r_e}{r_e - h^2} = 2$,

with solutions

$$\begin{bmatrix} \hat{b} \\ \hat{u}_A \\ \hat{u}_r \end{bmatrix} = \begin{bmatrix} C^{11} & C^{12} & C^{13} \\ C^{21} & C^{22} & C^{23} \\ C^{31} & C^{32} & C^{33} \end{bmatrix} \begin{bmatrix} X'y \\ Z'y \\ Z'y \end{bmatrix}$$

The terms C^{ii} denote the submatrices of the inverse of the left-hand side of the mixed model equations. For example, C^{22} is not equal to $\left(Z'Z+\lambda A^{-1}\right)^{-1}$. The C^{ii} submatrices are required for calculation of prediction error variances and covariances (Henderson, 1975).

Prediction error variances of the predicted breeding values are the diagonal terms of the C^{22} submatrix, multiplied by the general environmental variance, $\sigma_E^2 = (1-r_e)\sigma_P^2$.

Similarly, the prediction error variances for future performance of animals are equal to the diagonal terms of the sum of the matrices

$$\left(C^{22} + C^{33} + 2C^{23}\right)\sigma_E^2$$

In practice, with large numbers of animals, calculation of prediction error variances by matrix inversion is not always feasible, such that the diagonal elements of the inverse matrix are approximated (Thompson et al., 1994).

Solutions to the mixed model equations are as follows:

Animal (additive genetic)			Animal (specific environment)			Sire	Dam
1	2	3	1	2	3		
0.28	0.50	−0.58	0.08	0.31	−0.39	0.39	0.00

with estimates of the month effects equal to 9.81, 11.31 and 13.58.

The future performance of the three cows is predicted from the sum of the additive genetic and special environmental effects of each animal, which are 0.36, 0.81 and −0.97.

The prediction error variances for the predicted breeding values and for future performance of the three cows are

$$\begin{bmatrix} 3.35 \\ 3.50 \\ 3.17 \end{bmatrix} \quad \text{and} \quad \left\{ \begin{bmatrix} 0.84 \\ 0.88 \\ 0.79 \end{bmatrix} + \begin{bmatrix} 0.42 \\ 0.44 \\ 0.41 \end{bmatrix} + 2 \begin{bmatrix} -0.11 \\ -0.08 \\ -0.14 \end{bmatrix} \right\} \sigma_E^2$$

Prediction Error Variance

The additive genetic variance is the sum of the variance of the predicted breeding value and the prediction error variance, as discussed in Chapter 4. The variance of the predicted breeding value tends to the additive genetic variance as the accuracy of the predicted breeding value increases, such that the prediction error variance decreases. Solving the mixed model equations provides both the predicted breeding values and the prediction error variances, such that the accuracy of a predicted breeding value can be determined from the prediction error variance.

The effective number of records can be estimated, given the prediction error variance, which is an indication of the number of records on the animal, assuming that the records were evenly distributed between the different classes of the fixed effects. If there was a high degree of confounding between a fixed effect and an animal, then there would be a substantial difference between the actual number of records and the effective number of records. The effective number of records can give an indication of the balanced nature of the data.

The prediction error variance, PEV, is

$\text{PEV} = \left(1 - r_{A\hat{A}}^2\right)\sigma_A^2$ where $r_{A\hat{A}}$ is the accuracy of the predicted breeding value

$\text{PEV} = \dfrac{\lambda}{n_e + \lambda}\sigma_A^2$ where n_e is the effective number of records

$PEV = \sigma_A^2 - \sigma_{\hat{A}}^2$ where $\sigma_{\hat{A}}^2$ is the variance of the predicted breeding value

Interpretation of the prediction error variance is dependent on whether the breeding values are predicted with a sire model or with an individual animal model. In particular,

with a sire model $PEV = \text{diag}(C^{22})\sigma_e^2$ where $\sigma_s^2 + \sigma_e^2 = \sigma_P^2$

with an animal model $PEV = \text{diag}(C^{22})\sigma_E^2$ where $\sigma_A^2 + \sigma_E^2 = \sigma_P^2$

The accuracy of the predicted breeding value is

$$r_{A\hat{A}} = \sqrt{1 - \frac{PEV}{\sigma_s^2}}$$ with a sire model

and

$$r_{A\hat{A}} = \sqrt{1 - \frac{PEV}{\sigma_A^2}}$$ for an individual animal model

The effective number of records is

$$n_e = [PEV^{-1} - \sigma_s^{-2}]\sigma_e^2$$ with a sire model

$$n_e = [PEV^{-1} - \sigma_A^{-2}]\sigma_E^2$$ for an individual animal model

With a sire model, the interpretation of $\hat{u}$ is half the predicted breeding value. The variance of the predicted breeding value is $4\,\text{var}(\hat{u})$ and the prediction error variance of the predicted breeding value is $4PEV(\hat{u})$.

Chapter twelve

Multivariate Breeding Value Prediction

Information from several traits can be incorporated into the mixed model equations, for prediction of breeding values, in a manner analogous to predicting an animal's genetic merit for one trait using information on two traits as in a selection criterion. The mixed model equations for one trait,

$$\begin{bmatrix} X'X & X'Z \\ Z'X & Z'Z + \lambda A^{-1} \end{bmatrix} \begin{bmatrix} b \\ u \end{bmatrix} = \begin{bmatrix} X'y \\ Z'y \end{bmatrix}$$

where $\lambda = \dfrac{\sigma_e^2}{\sigma_u^2}$, can be extended for several traits (Henderson and Quaas, 1976).

Equal Design Matrices

When several traits are measured on all animals, the incidence matrix for fixed effects for trait i, X_i, will be the same as the incidence matrix for fixed effects of trait j, X_j, and similarly for the incidence matrices for the random effects, Z_i and Z_j. The incidence matrices are referred to as the design matrices, such that when all traits are recorded on all animals, the design matrices are equal.

For two traits, the design matrices for the fixed effects, X_1 and X_2, and the matrices for the random effects, Z_1 and Z_2, can be expressed as submatrices of the X and Z matrices

$$X = \begin{bmatrix} X_1 & 0 \\ 0 & X_2 \end{bmatrix} \quad \text{and} \quad Z = \begin{bmatrix} Z_1 & 0 \\ 0 & Z_2 \end{bmatrix}$$

Similarly, the vectors for the fixed, b_1 and b_2, and random, u_1 and u_2, effect solutions and the traits, y_1 and y_2, can be expressed as subvectors of the b, u and y vectors

$$b = \begin{bmatrix} b_1 \\ b_2 \end{bmatrix}, \quad u = \begin{bmatrix} u_1 \\ u_2 \end{bmatrix} \quad \text{and} \quad y = \begin{bmatrix} y_1 \\ y_2 \end{bmatrix}$$

The model for the two traits can be written as

$$\begin{bmatrix} y_1 \\ y_2 \end{bmatrix} = \begin{bmatrix} X_1 & 0 \\ 0 & X_2 \end{bmatrix} \begin{bmatrix} b_1 \\ b_2 \end{bmatrix} + \begin{bmatrix} Z_1 & 0 \\ 0 & Z_2 \end{bmatrix} \begin{bmatrix} u_1 \\ u_2 \end{bmatrix} + \begin{bmatrix} I_1 & 0 \\ 0 & I_2 \end{bmatrix} \begin{bmatrix} e_1 \\ e_2 \end{bmatrix}$$

which is analogous to the model for one trait

$$y = Xb + Zu + e$$

The (co)variance structures for the residuals and random effects are

$$\text{var}\begin{bmatrix} e_1 \\ e_2 \end{bmatrix} = \begin{bmatrix} I\sigma_{e_1}^2 & I\sigma_{e_1 e_2} \\ I\sigma_{e_1 e_2} & I\sigma_{e_2}^2 \end{bmatrix} = R \quad \text{and} \quad \text{var}\begin{bmatrix} u_1 \\ u_2 \end{bmatrix} = \begin{bmatrix} A\sigma_{u_1}^2 & A\sigma_{u_{12}} \\ A\sigma_{u_{12}} & A\sigma_{u_2}^2 \end{bmatrix} = G$$

where $\sigma_{e_i}^2$ and $\sigma_{u_i}^2$ are the residual and random effect variances for trait i, while $\sigma_{e_1 e_2}$ and $\sigma_{u_1 u_2}$ are the corresponding covariances. The matrices R and G are square matrices of size 2n, where n is the number of measured animals.

The mixed model equations for two traits are

$$\begin{bmatrix} X'R^{-1}X & X'R^{-1}Z \\ Z'R^{-1}X & Z'R^{-1}Z + G^{-1} \end{bmatrix} \begin{bmatrix} b \\ u \end{bmatrix} = \begin{bmatrix} X'R^{-1}y \\ Z'R^{-1}y \end{bmatrix}$$

The form of the mixed model equations for two traits is the same as for one trait. With one trait the matrix R^{-1} is equal to the matrix $\sigma_e^{-2}I$, and after multiplying throughout by σ_e^2, the mixed model equations for one trait are:

$$\begin{bmatrix} X'X & X'Z \\ Z'X & Z'Z + \lambda A^{-1} \end{bmatrix} \begin{bmatrix} b \\ u \end{bmatrix} = \begin{bmatrix} X'y \\ Z'y \end{bmatrix}$$

Predicted Genetic Merit with Unmeasured Traits

When breeding values are predicted for several measured traits, then an animal's predicted genetic merit is defined as

$$a' \hat{u} = a_1 \hat{u}_1 + a_2 \hat{u}_2 + ... + a_n \hat{u}_n$$

where a_i is the economic value and $\hat{u}_i$ is the predicted breeding value for trait i.

Traits included in the selection objective may be measured on some animals but not on others. For example, the selection objective may consist of growth rate and reproductive performance, with growth rate measured on both males and females. The genetic relationship between male and female sibs will enable breeding values for reproductive performance to be predicted for males. A second situation is when all animals have measurements for one set of traits, but the selection objective contains traits for which no animals have measurements. For example, the selection objective may consist of growth rate and carcass lean content, but animals have measurements on growth rate and ultrasonic backfat depth. A third situation is when some animals are performance tested in one of two test environments, where a trait, such as growth rate, would be treated as two traits, growth rate in one environment and growth rate in the second environment, and breeding values for the two traits could be predicted for all animals. However, such a situation is discussed in the section on geneotype with environment interaction, later in this chapter.

Information on the genetic covariances between measured traits and traits in the selection objective can be incorporated into the mixed model equations, for prediction of breeding values for all traits in the selection objective.

For example, if the selection objective is to improve traits 1 and 2, but only trait 1 is measured, then using the format of the mixed model equations for two traits

$$\begin{bmatrix} X'R^{-1}X & X'R^{-1}Z & 0 \\ Z'R^{-1}X & Z'R^{-1}Z+A^{-1}g^{11} & A^{-1}g^{12} \\ 0 & A^{-1}g^{21} & A^{-1}g^{22} \end{bmatrix} \begin{bmatrix} b \\ u_1 \\ u_2 \end{bmatrix} = \begin{bmatrix} X'R^{-1}y \\ Z'R^{-1}y \\ 0 \end{bmatrix}$$

where

$$G^{-1} = \begin{bmatrix} \sigma^2_{A1} & \sigma_{A12} \\ \sigma_{A21} & \sigma^2_{A2} \end{bmatrix}^{-1} = \begin{bmatrix} g^{11} & g^{12} \\ g^{21} & g^{22} \end{bmatrix}$$

Trait 2 is not included in the mixed model equations, since

$$g^{21}A^{-1}\hat{u}_1 + g^{22}A^{-1}\hat{u}_2 = 0$$

Therefore, the predicted breeding values for trait 2 are calculated directly from the predicted breeding values of traits 1, as

$$-\sigma_{A21}\hat{u}_1 + \sigma^2_{A1}\hat{u}_2 = 0$$

or

$$\hat{u}_2 = \frac{\sigma_{A21}}{\sigma^2_{A1}}\hat{u}_1$$

which is the genetic regression of trait 2 on trait 1.

Similarly, if traits 1,2...,n are measured and have breeding values $\hat{u}_1, \hat{u}_2..., \hat{u}_n$, but a different set of traits, i,j...,k, are included in the selection objective, then the predicted breeding values of traits i,j...,k are obtained from

$$\begin{bmatrix} \hat{u}_i \\ \hat{u}_j \\ \vdots \\ \hat{u}_k \end{bmatrix} = \begin{bmatrix} \sigma_{Ai1} & \sigma_{Ai2} & \cdots & \sigma_{Ain} \\ \sigma_{Aj1} & \sigma_{Aj2} & \cdots & \sigma_{Ajn} \\ \vdots & \vdots & & \vdots \\ \sigma_{Ak1} & \sigma_{Ak2} & \cdots & \sigma_{Akn} \end{bmatrix} \begin{bmatrix} \sigma^2_{A1} & \sigma_{A12} & \cdots & \sigma_{A1n} \\ \sigma_{A21} & \sigma^2_{A2} & \cdots & \sigma_{A2n} \\ \vdots & \vdots & & \vdots \\ \sigma_{An1} & \sigma_{An2} & \cdots & \sigma^2_{An} \end{bmatrix}^{-1} \begin{bmatrix} \hat{u}_1 \\ \hat{u}_2 \\ \vdots \\ \hat{u}_n \end{bmatrix}$$

Overall genetic merit is

$$a'\hat{u} = a_i\hat{u}_i + a_j\hat{u}_j + ... + a_k\hat{u}_k$$

Common Environmental Effect

Mixed model methodology can be used to predict the common environmental effect, which is responsible for increasing the similarity between full-sibs, as they share the same environment. If the common environmental variance component is σ_C^2 and $\gamma = \dfrac{\sigma_E^2}{\sigma_C^2}$, then using the format for the mixed model equations for two traits, the mixed model equations for a model including a common environmental effect are

$$\begin{bmatrix} X'X & X'Z_a & X'Z_c \\ Z_a'X & Z_a'Z_a + \lambda A^{-1} & Z_a'Z_c \\ Z_c'X & Z_c'Z_a & Z_c'Z_c + \gamma I \end{bmatrix} \begin{bmatrix} b \\ a \\ c \end{bmatrix} = \begin{bmatrix} X'y \\ Z_a'y \\ Z_c'y \end{bmatrix}$$

where Z_c is a square matrix with dimensions equal to the number of litters. An estimate of the proportion of the phenotypic variance due to the common environmental effect, $c^2 = \dfrac{\sigma_C^2}{\sigma_P^2}$, is required for accuracte prediction of breeding values. Further, the decision whether or not to record a trait on full-sibs will depend on the magnitude of the common environmental effect.

Example
In a study of Hereford cattle, the heritability of weaning weight was 0.26, but when a maternal (common) environmental effect was included in the model the heritability reduced to 0.10 and the maternal environmental effect was 0.29 σ_P^2.(Meyer, 1992). If the maternal environmental effect was ignored for breeding value prediction, then the breeding values would be overestimated.

Maternal Genetic Effect

The genotype of the dam affects the phenotype of her progeny through both the additive genetic effect and the maternal genetic effect. The maternal genetic effect is the influence of the dam's genotype for maternal effects which influence the trait of interest in her progeny. For example, larger dams may provide more milk to their offspring, such that progeny of heavier dams will be heavier than progeny of lighter dams, to a greater extent than expected from the dam's genotype for weight. Note that the common environmental effect is not the same as the maternal genetic effect, as the common environmental effect is independent of the dam's genotype for the measured trait.

Prediction of an animal's breeding value should take account of the maternal genetic effect, particularly if the maternal genetic effect is related to the additive genetic effect. Prediction of both the additive genetic and the maternal genetic effects can be considered as a form of multivariate analysis (Quaas and Pollak, 1980).

Mixed model equations for a model including a maternal genetic effect are

$$\begin{bmatrix} X'X & X'Z_a & X'Z_m \\ Z_a'X & Z_a'Z_a + \lambda_a A^{-1} & Z_a'Z_m + \lambda_{am} A^{-1} \\ Z_m'X & Z_m'Z_a + \lambda_{am} A^{-1} & Z_m'Z_m + \lambda_m A^{-1} \end{bmatrix} \begin{bmatrix} b \\ a \\ m \end{bmatrix} = \begin{bmatrix} X'y \\ Z_a'y \\ Z_m'y \end{bmatrix}$$

by again using the format for the mixed model equations for two traits, where, in this case, a and m represent the additive and maternal genetic effects. The variance–covariance matrix for the additive and maternal genetic effects is

$$\text{var}\begin{bmatrix} a \\ m \end{bmatrix} = \begin{bmatrix} A\sigma_A^2 & A\sigma_{AM} \\ A\sigma_{AM} & A\sigma_M^2 \end{bmatrix}$$

with

$$\begin{bmatrix} \lambda_a & \lambda_{am} \\ \lambda_{am} & \lambda_m \end{bmatrix} = \sigma_e^2 \begin{bmatrix} \sigma_A^2 & \sigma_{AM} \\ \sigma_{AM} & \sigma_M^2 \end{bmatrix}^{-1}$$

The Z_a and Z_m matrices correspond to the incidence matrices for the additive and maternal genetic effects, respectively, and are determined in a comparable manner to that discussed in Chapter 11.

Each animal will have a predicted breeding value and a predicted maternal genetic effect, even though some animals have no progeny. An analogous situation was discussed in Chapter 11, when the predicted breeding values and specific environmental effects could be determined for all animals with records, even though some animals had only one record. REML methodology can be used to estimate the additive and maternal genetic variance component, for estimation of the maternal heritability, $h_m^2 = \dfrac{\sigma_M^2}{\sigma_A^2 + \sigma_M^2 + \sigma_E^2}$, and the correlation between additive genetic and maternal genetic effects, $r_{AM} = \dfrac{\sigma_{AM}}{\sigma_A \sigma_M}$.

Example

Returning to the study of Hereford cattle, weaning weight had a heritability, h^2, of 0.14, a maternal genetic effect, h_m^2, of 0.13, and a maternal (common) environmental effect, c^2, of 0.23, with a correlation between additive genetic and maternal genetic effects of −0.59. The estimated parameters indicate that the genotype of the calf for weaning weight is less important than the maternal genetic and maternal environmental effects, since $h^2 < h_m^2 + c^2$ (Meyer, 1992). The negative correlation between additive genetic and maternal genetic effects for weaning weight suggests an antagonism between genes for post-natal growth, with genes controlling the maternal environment.

When the maternal genetic effect was not included in the model, there were no substantial changes in the other parameters ($h^2 = 0.10$ and $c^2 = 0.29$), as noted in the previous section, but when the maternal (common) environmental effect was not included in the model, there was no change in the heritability, $h^2 = 0.14$, but the maternal genetic effect increased, $h_m^2 = 0.46$. Clearly, breeding values predicted with an inappropriate model will be poor indicators of the performance of progeny of selected animals.

Unequal Design Matrices

Rather than being able to estimate genetic and phenotypic parameters only when all animals have measurements on all traits, with equal X and Z matrices for each trait, the REML procedure can also be used to estimate the genetic and phenotypic parameters when not all animals have measurements on all traits (Thompson *et al.*, 1995). For example, growth traits can be measured on males and females, but reproduction traits can only be measured on females. Information on growth traits measured on males can be incorporated in the analysis of both growth and reproduction traits on females.

Example
The genetic correlation between pig litter weight at birth was positive with growth rate, 0.33, but negative with food conversion ratio, −0.46, and backfat depth, −0.33, when estimated from data of farrowing Large White gilts from lines selected on performance test traits. When performance test data on all boars and all non-farrowing gilts was incorporated with performance test and farrowing data of gilts which farrowed, the estimated genetic correlations were 0.65, −0.23 and −0.29, respectively. Exclusion of information from boars and non-farrowing gilts from the analysis may have resulted in negatively biased genetic correlation estimates (Kerr and Cameron, 1996).

Genotype with Environment Interaction

A second example of unequal design matrices is the estimation of the genetic correlation, r_A, between a trait measured in one environment with the same trait measured in a second environment (Schaeffer *et al.*, 1978). The same trait in the two environments can be treated as two separate traits. Information on the genetic correlation between the trait measured in the two environments is required to determine whether the rate of genetic improvement in environment 1 would be higher with direct selection in environment 1 or with indirect selection in environment 2. The relative responses from selection in the two environments would be $\frac{r_A h_1 h_2}{h_1^2}$, where h_2^2 is the heritability of the trait in environment 2.

The genetic correlation is a measure of the relative ranking of genotypes in both environments and is a measure of the genotype with environment interaction, $G \times E$. Design of breeding programmes should take account of the $G \times E$, as the predicted relative response in environment 1 with selection in environment 2 may be substantially lower than the expected value of h_2/h_1, if no account is taken of the $G \times E$.

Estimation of the genetic correlation between traits measured in the two environments requires that the animals in the two environments are genetically related, with the relationship matrix for animals measured in environments 1 and 2 equal to A_{12}. Similarly, the relationship matrix for animals measured in environment 1 is A_{11}. The variance–covariance matrix for the random effects is

$$\text{var}\begin{bmatrix} u_1 \\ u_2 \end{bmatrix} = \begin{bmatrix} A_{11}\sigma_{u_1}^2 & A_{12}\sigma_{u_{12}} \\ A_{21}\sigma_{u_{12}} & A_{22}\sigma_{u_2}^2 \end{bmatrix} = G$$

There is no correlation between residuals in the two environments, as the traits are measured on different animals. The variance–covariance matrix for the residuals is

$$\text{var}\begin{bmatrix} e_1 \\ e_2 \end{bmatrix} = \begin{bmatrix} I\sigma_{e_1}^2 & 0 \\ 0 & I\sigma_{e_2}^2 \end{bmatrix} = \begin{bmatrix} I\gamma_1 & 0 \\ 0 & I\gamma_2 \end{bmatrix}^{-1}$$

Absorption of fixed effects

With multitrait analyses, the number of mixed model equations increases rapidly, as the dimensions of the matrices are proportional to the number of fixed and random effects to be estimated. One method to reduce the number of equations is to absorb the fixed effects, such that only equations for the random effects need to be solved. If the model is written as

$$Xb + Zu = y$$

then multiplying both sides of the equation by the transpose of X, multiplying by the inverse of (X'X) and then multiplying by X results in

$$Xb + X(X'X)^{-1}X'Zu = X(X'X)^{-1}X'y$$

Subtraction of the new equation from the initial equation removes the fixed effect terms, and the resulting equation is

$$Zu - X(X'X)^{-1}X'Zu = y - X(X'X)^{-1}X'y$$

If the matrix M is defined as $I - X(X'X)^{-1}X'$, then the above equation can be written as $MZu = My$ and the random effect solutions are obtained from

$$u = \left(Z'MZ + \lambda A^{-1}\right)^{-1} Z'My.$$

The mixed model equations for the G × E, after absorption of the fixed effects, are

$$\begin{bmatrix} \gamma_1 Z_1' M_1 Z_1 + G^{11} & G^{12} \\ G^{21} & \gamma_2 Z_2' M_2 Z_2 + G^{22} \end{bmatrix} \begin{bmatrix} u_1 \\ u_2 \end{bmatrix} = \begin{bmatrix} \gamma_1 Z_1' M_1 y_1 \\ \gamma_2 Z_2' M_2 y_2 \end{bmatrix}$$

or

$$\begin{bmatrix} C^{11} & C^{12} \\ C^{21} & C^{22} \end{bmatrix} \begin{bmatrix} u_1 \\ u_2 \end{bmatrix} = \begin{bmatrix} \gamma_1 Z_1' M_1 y_1 \\ \gamma_2 Z_2' M_2 y_2 \end{bmatrix}$$

where

$$G^{-1} = \begin{bmatrix} G^{11} & G^{12} \\ G^{21} & G^{22} \end{bmatrix}$$

Each animal will have two predicted breeding values, one for each environment, and the genetic variance for the trait in each environment can be estimated from the predicted breeding values, as for the univariate analyses, with the estimated genetic covariance for the trait in the two environments equal to

$$\sigma_{u_{12}} = \frac{\hat{u}_1' A^{12} \hat{u}_2 + tr\left[A^{12} C^{12}\right]}{n_{u_{12}}}$$

Example

In a population of Large White pigs, half the pigs within each litter were performance tested with *ad-libitum* feeding, with their littermates performance tested on restricted feeding. The genetic correlation between growth rate of *ad-libitum* fed pigs and growth rate of restricted fed pigs of 0.4 was significantly less than unity, which indicated a substantial genotype with feeding regime interaction. The genetic correlation for ultrasonic backfat depth was 0.8. When pigs selected for lean growth rate on restricted feeding were tested on *ad-libitum* feeding, they grew faster than pigs selected for lean growth rate on *ad-libitum* feeding. The genotype with feeding regime interaction for growth rate indicated that selection for lean growth rate should be based on restricted feeding, rather than on an *ad-libitum* feeding regime (Cameron and Curran, 1995).

Chapter thirteen

Breeding Values with a Gene of Known Large Effect

In previous chapters, the model has assumed that quantitative genetic variation is due to the effect of an infinite number of genes, each with small effect. The assumption is made to enable prediction of breeding values and responses to selection. The assumption is unrealistic, as there are a finite number of genes and single genes with large effects have been identified. For example, the oestrogen receptor gene in pigs (Rothschild *et al.*, 1996) and the Booroola gene in sheep (Piper and Bindon, 1985) are both associated with increased litter size. The ryanodine receptor (MacLennan *et al.*, 1990) or the "halothane" (Eikelenboom and Minkema, 1974) gene alters the susceptibility to porcine stress syndrome, there is a double muscling gene in cattle (Hanset and Michaux, 1985) and there are two genes associated with meat quality in pigs (Le Roy *et al.*, 1990; Janss *et al.*, 1994).

Although the assumptions of the infinitesimal model may be simplistic, the model appears to have been robust in terms of predicting responses to selection. If there are genes of known large effect, then incorporation of information on the effects of these genes is expected to increase the accuracy of predicted breeding values and estimated responses to selection. Secondly, DNA can be sampled from an animal at any age, such that the generation interval can be reduced if animals are selected directly on their genotype, rather than waiting for a phenotypic measurement.

Parameterising a Gene of Known Large Effect

For genes of known large effect, the genetic model assumes that the difference between the two homozygotes is equal to twice the effect of the favourable allele, denoted by a, and the difference between the heterozygote and the average of the two homozygotes is the heterozygote advantage, denoted by d. Quantifying the effect of a single gene requires information on homozygotes with two copies of the favourable allele, A_1A_1, on homozygotes with no copies of the favourable allele, A_2A_2, and the heterozygotes, which have a copy of each allele, A_1A_2. The effect of the favourable allele is

$$a = \frac{\overline{A_1A_1} - \overline{A_2A_2}}{2}$$

where $\overline{A_1A_1}$ is the mean phenotype of homozygotes with the favourable allele. Similarly, the heterozygote advantage is

$$d = \overline{A_1A_2} - \frac{\left(\overline{A_1A_1} + \overline{A_2A_2}\right)}{2}$$

which measures the difference between the observed value of the combination of the two alleles and the expected value of the combination of the two alleles, based on information from the homozygotes.

If the phenotypes of the homozygote with the favourable allele and the heterozygote are equal, then d equals a. When the heterozygote is equal to the average of the two homozygotes, there is no heterozygote advantage, d is zero and the gene is referred to as being completely additive.

For example, in a Merino population (Piper et al., 1985), mean ovulation rate and litter size of the three genotypes are as shown in Table 13.1.

Table 13.1. Ovulation rate and litter size of the three Merino genotypes

	Homozygote with Booroola gene	Heterozygote	Normal homozygote
Ovulation rate	4.3	2.8	1.3
Litter size	2.7	2.1	1.2
Model	$\mu + a$	$\mu + d$	$\mu - a$

The allele effect for ovulation rate is 1.5 and that for litter size is 0.75, while the heterozygote advantage is equal to 0 for ovulation rate and 0.15 for litter size.

Genotypic Values and Breeding Values

In previous chapters, breeding values were determined from the regression of genotype on phenotype. For a gene of known large effect with the frequency of the favourable allele equal to p, the regression coefficient is estimated from a weighted regression of the assigned values to each genotype (a, d and −a) on the number of A_1 alleles (2, 1 and 0), with the genotype frequencies as weights (p^2, 2pq and q^2), where $q = 1 - p$. The genotypic values are the difference between the assigned values and the mean genetic merit of the population, with the latter equal to

$$ap^2 + d2pq - aq^2 = a(p-q) + 2dpq$$

Breeding values for the gene of known large effect are estimated by multiplying the genotypic values by the regression coefficient. The difference between the genotypic value and the breeding value is equal to the dominance deviation, which accounts for the non-additive components of genetic variation (see Table 13.2).

Table 13.2. Formulae for estimating genotypic and breeding values

	Genotype		
	A_1A_1	A_1A_2	A_2A_2
Assigned value	a	d	$-$a
Frequency	p^2	$2pq$	q^2
Genotypic value	$2q(a-dp)$	$a(q-p)+d(1-2pq)$	$-2p(a+dq)$
Breeding value	$2q\alpha$	$(q-p)\alpha$	$-2p\alpha$
Dominance deviation	$-2dq^2$	$2dpq$	$-2dp^2$

$\alpha = a + d(q-p)$.

If the frequency of the favourable allele is 0.5, then α is equal to a and the mean genetic merit of the population is d/2. The breeding values of the three genotypes are a, 0 and $-$a, and the sums of the genotypic values plus the mean genetic merit of the population are a, d and $-$a, respectively.

The additive genetic variance accounted for by the gene of known large effect is the sum of the squared breeding values multiplied by the genotype frequencies:

$$(2q\alpha)^2 p^2 + (q-p)^2 \alpha^2 2pq + (-2p\alpha)^2 q^2 = 2pq\alpha^2$$

Similarly, the non-additive genetic variance is $(2dpq)^2$

The total variance accounted for by the gene of known large effect, σ_G^2, is

$$2pq(\alpha^2 + 2d^2pq)$$

The phenotypic variance of the trait, σ_P^2, is $\sigma_G^2 + \sigma_A^2 + \sigma_E^2$, where the polygenic variance and environmental variance are σ_A^2 and σ_E^2, respectively.

Further information on the variance formula is given by Falconer and MacKay (1996).

Example

A trait with a completely additive gene of known large effect has a phenotypic variance of 1, a total genetic variance, $\sigma_G^2 + \sigma_A^2$, of 0.25, of which the polygenic variance, σ_A^2, is 0.20 and the frequency of the favourable allele, p, is 0.15. The genotypic values equal the breeding values and the three genotypes are determined from $2pq(\alpha^2 + 2d^2pq) = \sigma_G^2 = 0.05$ and d=0:

	Genotype		
	A_1A_1	A_1A_2	A_2A_2
Assigned value	0.443	0	$-$0.443
Frequency (%)	2.25	25.5	72.25
Genotypic value	0.75	0.31	$-$0.13

Selection with a Gene of Known Large Effect

Assuming that the gene of large effect has been identified, the effect of the gene is known and all animals can be genotyped without error, then the phenotype of an animal can be expressed as

$$P = \frac{G_S + G_D}{2} + G_{MS} + (A + E)$$

where $\frac{G_S + G_D}{2}$ is the average genotypic value for the gene of known large effect of the animal's sire and dam. The term G_{MS} is the Mendelian sampling term for the gene of known large effect, A is the polygenic effect and E is the environmental effect. The polygenic and environmental effects are combined for the purpose of selection, which is exactly the same as when breeding values were predicted under the assumption of the infinitesimal model.

An animal's breeding value can be predicted from the three components of the animal's phenotype (Woolliams and Pong–Wong, 1995), which can be considered as three traits in a selection criteria, when the selection objective is to improve the overall genotype. The general selection criterion is

$$I = b_1 \frac{G_S + G_D}{2} + b_2 G_{MS} + b_3 (A + E)$$

The P and G matrices and the vector of economic values, a, required for calculation of the selection criterion coefficients are

$$P = \begin{bmatrix} \frac{1}{2}\sigma_G^2 & 0 & 0 \\ 0 & \frac{1}{2}\sigma_G^2 & 0 \\ 0 & 0 & \sigma_A^2 + \sigma_E^2 \end{bmatrix}, \quad G = \begin{bmatrix} \frac{1}{2}\sigma_G^2 & 0 & 0 \\ 0 & \frac{1}{2}\sigma_G^2 & 0 \\ 0 & 0 & \sigma_A^2 \end{bmatrix}, \quad a = \begin{bmatrix} 1 \\ 1 \\ 1 \end{bmatrix}$$

The genotypic effect and the Mendelian sampling term have a variance of $\frac{1}{2}\sigma_G^2$, as the genetic variance attributable to the gene of known large effect, σ_G^2, is the sum of the genotypic effect and the Mendelian sampling term. Genetic covariances between the genotypic effect, the Mendelian sampling term and the polygenic effect are assumed to be zero. Similarly, covariances between genetic effects and the environmental effect are also assumed to be zero. Economic values for the genotypic effect, the Mendelian sampling term and the polygenic effect are equal, and arbitrarily set to one, as the selection objective is to improve the overall genetic merit.

The selection criterion coefficients for the genotypic effect, the Mendelian sampling term and the polygenic effect, which maximise the accuracy of the predicted breeding value, are 1, 1 and $h_P^2 = \sigma_A^2 / (\sigma_A^2 + \sigma_E^2)$, respectively, where h_P^2 is the heritability of the polygenic component of the trait, after accounting for the gene of known large effect.

An alternative selection criterion, consisting of the Mendelian sampling term and the polygenic effect, would reduce the emphasis on the mean family genetic merit. Selection on the Mendelian sampling term is analogous to selection on the within-family deviation, as discussed in Chapter 5, such that family members are less likely to be selected, compared to when the selection criterion includes the genotypic effect, with a corresponding reduction in the rate of inbreeding. The selection criterion coefficients can either be obtained using the P and G matrices for selection on the genotypic effect, the Mendelian sampling term and the polygenic effect, with the economic value for the genotypic effect equal to zero, or by using submatrices of the P and G matrices corresponding to the Mendelian sampling term and the polygenic effect. In either case, the selection criterion coefficients are 1 and h_P^2.

Finally, the third alternative selection criterion is phenotypic selection, which does not use information on the animal's genotype, and the selection criterion coefficients for the genotypic effect, the Mendelian sampling term and the polygenic effect are all equal to $h^2 = \left(\frac{1}{2}\sigma_G^2 + \sigma_A^2\right)/\sigma_P^2$, where the heritability is estimated assuming that there is no gene of large effect.

In summary, the three alternative selection criteria of genetic, Mendelian and phenotypic selection emphasise different components of the phenotype, as shown in Table 13.3.

Table 13.3. Selection criterion coefficients for genotypic, Mendelian and phenotypic selection

Selection	Value of selection criterion coefficient		
	b_1	b_2	b_3
Genotypic	1	1	h_P^2
Mendelian	0	1	h_P^2
Phenotypic	h^2	h^2	h^2

Example
Selection criterion coefficients and the accuracy of the predicted breeding value for genotypic, Mendelian and phenotypic selection on a trait with a completely additive gene of known large effect, a phenotypic variance of 1, a total genetic variance, $\sigma_G^2 + \sigma_A^2$, of 0.25, of which the polygenic variance, σ_A^2, is 0.20, and the frequency of the favourable allele, p, is 0.15, are as follows:

Selection	Value of selection criterion coefficient			Accuracy
	b_1	b_2	b_3	
Genotypic	1	1	0.210	0.607
Mendelian	0	1	0.210	0.546
Phenotypic	0.231	0.231	0.231	0.480

The higher accuracy of the predicted breeding value with genotypic selection compared to phenotypic selection illustrates the advantage of including both the genotypic effect and the Mendelian sampling term in the selection criterion. In the example, Mendelian selection has intermediate accuracy, compared with genotypic and phenotypic selection, since genotypic information is not included in the selection criterion.

Selection on genetic and polygenic effects versus genetic effect and phenotype

For the purposes of selection, the genotypic effect and Mendelian sampling term could be combined, to form the genetic effect, with the selection criterion consisting of the genetic effect and the polygenic effect. From the previous section, the selection criterion coefficients would be 1 and h_P^2 for the genetic effect and polygenic effect, respectively,

$$I = G_L + h_P^2(A + E)$$

where G_L represents the gene of known large effect.

Alternatively, the selection criterion could consist of the genetic effect and the phenotype:

$$I = b_1 G_L + b_2(G_L + A + E)$$

The P and G matrices and the vector of economic values, a, required to calculate the selection criterion coefficients, are

$$P = \begin{bmatrix} \sigma_G^2 & \sigma_G^2 \\ \sigma_G^2 & \sigma_P^2 \end{bmatrix}, \quad G = \begin{bmatrix} \sigma_G^2 & \sigma_G^2 \\ \sigma_G^2 & \sigma_A^2 + \sigma_G^2 \end{bmatrix}, \quad a = \begin{bmatrix} 0 \\ 1 \end{bmatrix}$$

The covariance between phenotype and gene effect is σ_G^2 and the economic value of the gene effect is zero, as both the genetic effect and polygenic effect are included in the phenotype. The selection criterion coefficients are

$$\frac{1}{\sigma_A^2 + \sigma_E^2} \begin{bmatrix} \sigma_E^2 \\ \sigma_A^2 \end{bmatrix} = \begin{bmatrix} 1 - h_P^2 \\ h_P^2 \end{bmatrix}$$

The selection criterion is

$$I = (1 - h_P^2)G_L + h_P^2(G_L + A + E)$$

or

$$I = G_L + h_P^2(A + E)$$

The two selection criterion are expected to be equal, as the same information is included in both criteria, although it is combined differently.

Responses to Selection

Responses to genotypic, Mendelian or phenotypic selection are dependent on the number of generations of selection, the frequency of the favourable allele, p, the proportion of the total variance accounted for by the gene of known large effect and the heritability of the polygenic component of the trait.

Example
Cumulative responses to genotypic, Mendelian and phenotypic selection are illustrated in Fig. 13.1 for several generations of selection on a trait with a completely additive gene of known large effect, a phenotypic variance of 1, and a total genetic variance, $\sigma_G^2 + \sigma_A^2$, of 0.25, of which the polygenic variance, σ_A^2, is 0.20 and the frequency of the favourable allele, p, is 0.15. Each generation, 20 males are selected from 180 and 60 females are selected from 180, with hierarchical mating at random (Woolliams and Pong-Wong, 1995).

Cumulative responses for genotypic or Mendelian selection are greater than for phenotypic selection only in the first five generations of selection. Genotypic selection results in a greater cumulative response than Mendelian selection in generations two and three, after which the cumulative response of Mendelian selection is marginally greater

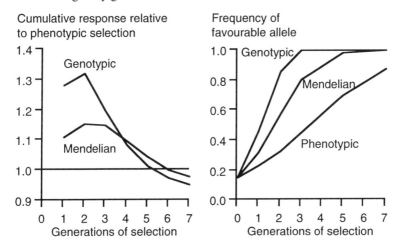

Fig. 13.1. Cumulative responses with genotypic or Mendelian selection and frequencies of the favourable allele

In the early generations of selection, the greater cumulative response with genotypic or Mendelian selection, relative to phenotypic selection, is due to a greater accuracy of predicted breeding value, resulting in a higher rate of increase in the frequency of the favourable allele (see Fig. 13.1 for the example). However, in the longer term, the gene of known large effect will be fixed (i.e. the frequency of the favourable allele equals unity) with genotypic or Mendelian selection, and the cumulative response will be lower than for phenotypic selection, due to relatively less emphasis on selection at the polygenic level.

In the example, three generations of genotypic selection are required before the frequency of the favourable allele is at least 0.99, five generations for Mendelian selection, but 12 generations for phenotypic selection. Therefore, after five generations of selection, sustained selection on both the gene of known large effect and the polygenic level by phenotypic selection results in a greater cumulative response than either genotypic or Mendelian selection.

Genetic Markers

In the previous section, it was assumed that the gene of large effect was known and that animals could be genotyped for the gene. If a gene of large effect exists, but the gene has not been identified, then direct selection on the genetic effect is not possible. However, sections of DNA surrounding the gene are likely to segregate with the gene, particularly if the distance from the gene is relatively small. If there is genetic variation in the section of DNA close to the gene and animals can be genotyped for the section of DNA, then animals could be selected on the basis of both the DNA segment and the phenotype. The small sections of DNA are referred to as genetic markers and incorporation of information from genetic markers in the selection criterion is called marker assisted selection. If the trait of interest is associated with several close genes that generally segregate together, then the group of genes is called a quantitative trait locus, QTL. There is essentially no difference between using marker assisted selection to predict genetic merit for a trait associated with a single gene or a QTL.

Detailed discussion of methods to identify markers for QTL is outwith the scope of this text, but the reader is referred to Lander and Botstein (1989). However, two simple, but not necessarily powerful, methods for identifying markers associated with QTL are illustrated.

Two inbred lines have different alleles at both the marker (M_1 and M_2) and the QTL (QTL_1 and QTL_2) and the performance of lines 1 and 2 is $\mu + a$ and $\mu - a$, respectively:

$$
\begin{array}{ccc}
\text{Line 1} & & \text{Line 2} \\
\dfrac{M_1 \;\; QTL_1}{M_1 \;\; QTL_1} & \times & \dfrac{M_2 \;\; QTL_2}{M_2 \;\; QTL_2}
\end{array}
$$

$$
\begin{array}{ccc}
\text{Line 1} & & \\
\dfrac{M_1 \;\; QTL_1}{M_1 \;\; QTL_1} & \times & \dfrac{M_1 \;\; QTL_1}{M_2 \;\; QTL_2}
\end{array}
$$

If the two lines are crossed and the F_1 cross is mated to line 1, then all progeny will inherit $M_1 \; QTL_1$ from line 1, but there are four possibilities for the F_1 cross, due to recombination, which has probability r, and the expected performance of the four possible genotypes is as shown in Table 13.4.

Table 13.4. Expected frequency and performance of genotypes from the backcross

Genotype	Frequency	Performance
$M_1 \; QTL_1 \; M_1 \; QTL_1$	$(1-r)/2$	$\mu + a$
$M_2 \; QTL_2 \; M_1 \; QTL_1$	$(1-r)/2$	$\mu + d$
$M_1 \; QTL_2 \; M_1 \; QTL_1$	$r/2$	$\mu + d$
$M_2 \; QTL_1 \; M_1 \; QTL_1$	$r/2$	$\mu + a$

The expected performance of M_1M_1 and M_1M_2 animals is $\mu + (1-r)a + rd$ and $\mu + ra + (1-r)d$, such that the difference in performance between animals with the M_1 and M_2 marker alleles is $(a-d)(1-2r)$. The difference between marker genotypes could be due to a gene with a small effect and a small distance from the marker, such that the recombination rate is low, or to a gene with a large effect and a large distance from the marker, such that the recombination rate is high. For example, the difference between marker genotypes is the same for a = 2, d = 1 and r = 0.1 as for a = 5, d = 1 and r = 0.4.

If the F_1 cross is mated to line 2, then the difference in performance between animals with the M_1 and M_2 marker alleles is $(a+d)(1-2r)$, rather than $(a-d)(1-2r)$. If the marker and QTL are completely linked, then the $M_1 - M_2$ difference will be $a+d$ when estimated by crossing F_1 with line 2, and equal to $a-d$ when estimated by crossing F_1 with line 1.

In an outbred population, differences between marker genotypes must be determined on a within-family basis, due to possible between-family differences in the linkage of the marker and the QTL. For example, linkage between the marker and the QTL is opposite for sires A and B, while there is no detectable association between the marker and the QTL for sires C and D:

```
    Sire A            Sire B            Sire C            Sire D
  M₁  QTL₁          M₁  QTL₂          M₁  QTL₁          M₂  QTL₁
  ──────────        ──────────        ──────────        ──────────
  M₂  QTL₂          M₂  QTL₁          M₂  QTL₁          M₂  QTL₂
```

The within-sire family association between the marker and the QTL can be determined by either measuring the performance of progeny and genotyping progeny for the marker, the daughter design, or by measuring the performance of grand-progeny and genotyping progeny for the marker, the grand-daughter design (Weller et al., 1990). The daughter and grand-daughter designs could be used in dairy cattle populations. In the grand-daughter design, performance would be measured on progeny of sons of the sire, the grand-progeny, to identify associations between the marker and the sons' predicted breeding values, based on their daughters' performance.

Breeding Values with Marker Assisted Selection

Incorporation of marker information in breeding value prediction exploits within-family linkage disequilibrium to increase the accuracy of the predicted breeding value. Linkage information between the marker and QTL alleles is incorporated in the mixed model equations, on a within-family basis only, as the association between a marker allele and the favourable QTL allele will be family dependent.

For a QTL, each animal has two alleles, one from each parent, and associated with the QTL is a genetic marker. An animal's predicted breeding value is

$$\hat{a} = u + v_S + v_D$$

where u is the polygenic effect, while v_S and v_D are the additive genetic effects associated with the markers from the sire and dam, respectively.

The model can be written as

$$y = Xb + Zu + W_S v_S + W_D v_D + e$$

or

$$y = Xb + Zu + Wv + e$$

where b and u are the vectors of fixed and polygenic effects, with corresponding incidence matrices, X and Z, with rows of the W matrix describing the presence or absence of the genetic marker from each parent (Fernando and Grossman, 1989; van Arendonk et al.,1994).

The mixed model equations are

$$\begin{bmatrix} X'X & X'Z & X'W \\ Z'X & Z'Z + \lambda A^{-1} & Z'W \\ W'X & W'Z & W'W + \gamma G_{v|r}^{-1} \end{bmatrix} \begin{bmatrix} b \\ u \\ v \end{bmatrix} = \begin{bmatrix} X'y \\ Z'y \\ W'y \end{bmatrix}$$

with $\lambda = \sigma_E^2/\sigma_A^2$ and $\gamma = \sigma_E^2/\sigma_V^2$, where σ_A^2 and σ_V^2 are the variances of the polygenic effect and the additive genetic effect associated with the marker. The total genetic variance is $\sigma_A^2 = \sigma_U^2 + 2\sigma_V^2$. The matrix $G_{v|r}$ is the variance–covariance matrix of the additive genetic effects associated with the markers from each animal's sire and dam, given the recombination rate.

Example

The pedigree from van Arendonk et al. (1994) is used to illustrate construction of the $G_{v|r}$ matrix. The marker genotype of each animal is shown in parentheses, with the first marker allele being the paternal allele. Animals 1 and 2 are unrelated:

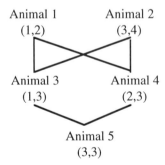

For animals 3, 4 and 5, the lower triangle of the numerator relationship matrix is as follows:

Animal					
3	0.5	0.5	1		
4	0.5	0.5	0.5	1	
5	0.5	0.5	0.75	0.75	1

Breeding values with a gene of large effect

If no account is taken of the marker genotypes, then the lower triangle of the relationship matrix for an animal's paternal (p) and maternal (m) QTL alleles with another animal's parents' paternal and maternal QTL alleles is as follows:

		\multicolumn{10}{c}{Animal}
		1		2		3		4		5	
Animal		p	m	p	m	p	m	p	m	p	m
3	p	0.5	0.5	0	0	1					
	m	0	0	0.5	0.5	0	1				
4	p	0.5	0.5	0	0	0.5	0	1			
	m	0	0	0.5	0.5	0	0.5	0	1		
5	p	0.25	0.25	0.25	0.25	0.5	0.5	0.25	0.25	1	
	m	0.25	0.25	0.25	0.25	0.25	0.25	0.5	0.5	0	1

Note that half the sum of terms for each animal is equal to the corresponding term in the numerator relationship matrix, A. For example, the probability that the paternally derived allele of animal 5 is the same as the paternally derived allele of animal 4 is 0.25, equal to $2 \times \left(\frac{1}{2}^2 \times \frac{1}{2}\right)$.

If the marker genotypes are taken into account, then the lower triangle of the relationship matrix for an animal's paternal (p) and maternal (m) QTL alleles with another animal's paternal and maternal QTL alleles is as follows:

		\multicolumn{10}{c}{Animal}
		1		2		3		4		5	
Animal		p	m	p	m	p	m	p	m	p	m
3	p	1−r	r	0	0	1					
	m	0	0	1−r	r	0	1				
4	p	r	1−r	0	0	2r(1−r)	0	1			
	m	0	0	1−r	r	0	A	0	1		
5	p	r(1−r)	r^2	$(1-r)^2$	r(1−r)	r	(1−r)	B	C	1	
	m	r^2	(1−r)r	$(1-r)^2$	r(1−r)	B	C	r	(1−r)	D	1

where $A = r^2 + (1-r)^2$, $B = 2r^2(1-r)$ and $C = r^2(1-r) + (1-r)^3$.

Elements of the relationship matrix are essentially the probability that the animal's paternal or maternal QTL allele are the same as another animal's paternal or maternal QTL allele, given the marker genotypes.

The relationship matrix also contains the inbreeding coefficient for both the marker and QTL alleles. For example, animal 5 is totally inbred for the marker allele, and assuming a recombination rate of 0.1, the inbreeding coefficient for the QTL is 0.67, equal to $D = (1-r)^4 + r^2(1-r)^2 + 2r^3(1-r)$, the probability that the paternal and maternal alleles are identical by descent.

The advantage of accounting for the marker genotype is illustrated with animal 5, as the relationship with animal 1 is substantially lower than with animal 2, compared to equal relationships when no account was taken of the marker:

	Account of marker				No account of marker			
	Animal 1		Animal 2		Animal 1		Animal 2	
	p	m	p	m	p	m	p	m
p	0.09	0.01	0.81	0.09	0.25	0.25	0.25	0.25
m	0.01	0.09	0.81	0.09	0.25	0.25	0.25	0.25

As the distance between a genetic marker and a QTL increases, the probability of recombination increases, which reduces the value of marker information. If two markers flank a QTL, the probability of mis-classifying an animal's QTL genotype on the basis of two genetic markers, r^2, is substantially lower than the probability of mis-classification given one marker, assuming a negligible probability of double recombination. Methodology for determining the relationship matrix for QTL alleles, based on information from two flanking markers, has been developed by Goddard (1992).

Simulation studies for dairy cattle (Meuwissen and van Arendonk, 1992), pig (Meuwissen and Goddard, 1996) and poultry (van der Beek and van Arendonk, 1996) breeding programmes have suggested that marker assisted selection could increase short terms rates of genetic response in nucleus breeding schemes by 10–20%, particularly when animals have to be selected before the trait can be measured, as in the case of carcass traits or longevity. However, as with direct selection on a gene of known large effect, the greater short-term response with marker assisted selection, compared to phenotypic selection, is not sustained in the long term.

Chapter fourteen
Breeding Values for Binary Traits

Several traits of interest in animal breeding, such as twinning in cattle or an animal contracting a disease, are of a binary nature. One of the problems with a binary trait, with phenotypic scores of zero and one, is that the mean value of the trait, p, is related to its variance $p(1-p)$, where p is the incidence of the phenotype scored as one. If the incidence of a particular phenotype depends on the group of animals observed, then the assumption of equal within-group variances will be inappropriate when variance components are estimated with analysis of variance techniques. Methods for estimation of variance components for binary traits have been developed, based on the concept of a threshold model.

Threshold Model and Liability

The threshold model assumes that phenotypic expression of the binary trait is determined by an underlying normally distributed trait, and that categorisation of the phenotype is defined by the value of the underlying trait. For example, if the underlying trait was related to an immune response which determined the animal's susceptibility to a disease, then values of the underlying trait below the

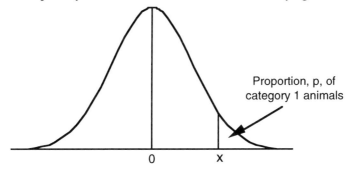

Fig. 14.1. Liability with the threshold model

threshold would result in the animal contracting the disease, while an animal would not contract the disease when the value of the underlying trait was above the threshold. The underlying normally distributed trait on a continuous scale with unit variance is called the liability. The value of the liability, x, which is assumed to differentiate between animals with phenotypes zero and one, is related to the incidence, p, of the binary trait, as illustrated in Fig. 14.1. The mean value of liability for category one animals is i, as described in Chapter 4.

For a normally distributed trait, the response to selection is predicted from the selection differential and the heritability of the trait. The selection differential is the difference between the phenotypic mean of the selected individuals and the population mean. With a binary trait, all animals in a particular category will have the same phenotype, but will have different liabilities. If the proportion of animals selected as parents is less than the incidence of the binary trait, then the selection differential for the liability is i, irrespective of the proportion selected, as animals are essentially selected at random from the category one animals. If the proportion of selected animals, s, is greater than the incidence of the binary trait, p, then the selection differential for liability is

$$\frac{ip(1-s)}{s(1-p)}$$

When the proportion of selected animals is equal to the incidence of the binary trait, then s equals p, and the selection differential for liability is i, as before.

The heritability of liability, h_L^2, can be estimated from the heritability of the binary trait, h_{01}^2:

$$h_L^2 = \frac{p(1-p)}{z^2} h_{01}^2 = \frac{(1-p)}{i^2 p} h_{01}^2$$

where z is the height of the normal distribution curve corresponding to the proportion p, and z equals ip (Robertson and Lerner, 1949). An estimate of the heritability of liability can be obtained by transforming the estimated heritability for the binary trait, which was determined using REML or analysis of variance with observations on progeny or from offspring–parent regression. In the formula for the heritability of liability, the numerator, $p(1-p)$ reflects the change from a normally distributed trait, with unit variance, to a binary trait with variance $p(1-p)$, and the denominator is an approximation to the non-linear relationship between liability and incidence.

The heritability of liability, h_L^2, is greater than the heritability of the binary trait, h_{01}^2, with their ratio increasing with decreasing incidence of the binary trait. For example, when the incidence is 0.5 and 0.1, the heritability for liability is 1.57 and 2.92 times greater than the heritability of the binary trait.

Response to Selection

An estimate of the predicted response to selection in a binary trait can be derived from the selection index methodology discussed in Chapter 6. The response in the binary trait, denoted by R_{01}, to selection is

$$R_{01} = i_{01} r_{A\hat{A}(01)} \sigma_{A(01)}$$

where i_{01} is the standardised selection differential for the binary trait, $r_{A\hat{A}(01)}$ is the accuracy of selection and $\sigma^2_{A(01)}$ is the genetic variance for the binary trait. If there are n repeated measurements on an animal for the binary trait, which has a repeatability of $r_{e(01)}$ and the animal is selected on the basis of the mean measurement, then the accuracy of the selection criterion is

$$r_{A\hat{A}(01)} = \sqrt{\frac{nh^2_{01}}{1+(n-1)r_{e(01)}}}$$

The binary trait has a phenotypic variance of $p(1-p)$. Therefore, the response to selection is

$$R_{01} = i_{01} h^2_{01} \sqrt{p(1-p)} \sqrt{\frac{n}{1+(n-1)r_{e(01)}}}$$

With one measurement of the binary trait, the response is

$$R_{01} = i_{01} h^2_{01} \sqrt{p(1-p)}$$

Note that the formula for the response with n measurements of the binary traits has the same form as the response to selection, on the basis of n repeated measurements of a normally distributed trait, as discussed in Chapter 4:

$$R = ih^2 \sigma_P \sqrt{\frac{n}{1+(n-1)r_e}}$$

Rather than directly predicting the response in the binary trait, it may be more appropriate to predict the response in the liability, a normally distributed trait, and then transform to the binary scale to predict the response in the binary trait. The response in the liability is

$$R_L = i_{01} r_{A\hat{A}(L)} h_L$$

where i_{01} is the standardised selection differential for the binary trait, $r_{A\hat{A}(L)}$ is the accuracy of selection on the liability and the genetic variance of the liability is h^2_L. If there are n repeated measurements on an animal for the binary trait, then the accuracy of the selection criterion (see Foulley, 1992) is

$$r_{A\hat{A}(L)} = r_{A\hat{A}(01)} = \sqrt{\frac{nh_{01}^2}{1+(n-1)r_{e(01)}}}$$

The predicted response in the liability is

$$R_L = i_{01}h_Lh_{01}\sqrt{\frac{n}{1+(n-1)r_{e(01)}}}$$

If animals are selected on the basis of one measurement of the binary trait, then the response in the liability is

$$R_L = i_{01}h_Lh_{01}$$

The response in liability is the difference between the value of the liability in the parental generation, x_{par}, which differentiates between animals in categories zero and one, and the liability threshold of the progeny generation, x_{prog}:

$$R_L = i_{01}h_Lh_{01} = x_{par} - x_{prog}$$

The expected incidence of the trait in the progeny generation is determined from the value of x_{prog},

$$x_{prog} = x_{par} - i_{01}h_Lh_{01}$$

and the response in incidence is the difference between the incidences of the progeny and parental generations.

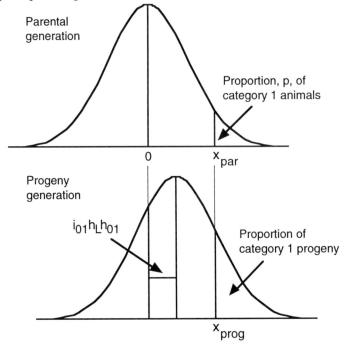

Fig. 14.2. Response in liability and incidence to selection

Example

The heritability and incidence of a binary trait are 0.15 and 0.1, respectively, and all animals in category one are selected as parents. The predicted response in the binary trait is 0.079, as the standardised selection differential corresponding to a selected proportion of 0.1 is 1.755.

The heritability of the liability is 0.438 and the response in liability is 0.450. The liability thresholds in the parental and progeny generations are 1.282 and 0.832, with a corresponding incidence in the progeny generation of 0.20, such that the response in the binary trait is 0.103.

The response in incidence depends on the heritability and the incidence of the binary trait as illustrated in Fig. 14.3, where the proportion of selected parents is 0.10 for all values of the incidence. The response in incidence is more dependent on the heritability of the binary trait than on the incidence, particularly when the incidence is at least 0.2. In general, the predicted response in the binary trait was greater when derived from the predicted response in the liability than when predicted directly. The difference in the predicted responses using the two methods increased as the heritability of the binary trait increased and as the incidence of the binary trait decreased.

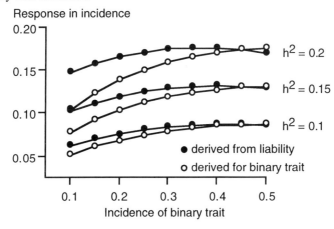

Fig. 14.3. Response in incidence, given a selection proportion of 0.10, with responses derived for the binary trait or from the liability, with different heritabilities of the binary trait

Methods of predicting the response in a binary trait with selection based on progeny measurements was discussed by Foulley (1992).

Heritability Estimation

The heritability of liability for a categorical trait with m levels can be determined from the heritability of the categorical trait, h^2_{cat}:

$$h_L^2 = h_{cat}^2 \left[\sum_{j=1}^{m} i_j^2 p_j - \left(\sum_{j=1}^{m} i_j p_j \right)^2 \right] / \left[\sum_{j=1}^{m-1} z_j \left(i_{j+1} - i_j \right) \right]^2$$

where category j has an incidence of p_j and a mean value of i_j, and z_j is the height of the normal distribution curve corresponding to the proportion p_j (Gianola, 1982). The mean values of each category are determined from standardised normal deviates of the normal distribution, corresponding to the proportion of animals in each category.

The standardised normal deviate for the extreme categories can be determined directly from normal distribution tables. For intermediate category k, the standardised normal deviate, i_k, is determined from

$$p_k i_k = (p_j + p_k) i_{jk} - p_j i_j$$

where p_j is the proportion of observations in more extreme categories with corresponding normal deviate i_j, p_k is the proportion of observations in category k and i_{jk} is the normal deviate corresponding to proportion $(p_j + p_k)$. If the actual category scores are replaced by the standardised normal deviates, then the categorical trait will have a mean and variance of 0 and 1, respectively.

For example, an eating quality trait has five scores of 1, 2..., 5, and the proportions of animals in each category are 0.40, 0.25, 0.20 0.10 and 0.05, respectively. The two extreme categories, of 1 and 5, have standardised normal deviates, i_1 and i_5, of −0.966 and 2.063, with the corresponding standardised normal deviates for categories 2, 3 and 4 equal to −0.064, 0.686 and 1.300, respectively. The normal deviates for the five categories are not necessarily equally spaced, as the differences between adjacent categories are 0.90, 0.75, 0.61 and 0.76. Therefore, under the assumption that there is an underlying trait with a normal distribution, the difference between eating quality scores 1 and 2 is not the same as the difference between scores 3 and 4.

For a binary trait, the formula for deriving the heritability of liability from the heritability of the categorical trait reduces to

$$h_L^2 = \frac{p(1-p)}{z^2} h_{01}^2$$

as the mean values for the two classes are $-ip/(1-p)$ and i, respectively.

The transformation of the heritability for the binary trait to the heritability of liability is only approximate, as the derived heritability of liability is generally overestimated. The positive bias increases as the heritability for the binary trait increases, particularly at low incidences of the trait. The magnitude of the bias can be demonstrated empirically. The generation of a continuous trait and a corresponding binary trait, derived by imposing a threshold point on the continuous trait, enables comparison of the actual heritability of liability, h_L^2, with the derived heritability for liability, based on the heritability of the binary trait, h_{01}^2. For example, when the heritability of liability, h_L^2, is 0.10, 0.25 or 0.40 and the incidence is 0.2, the derived heritability of liability is 0.10, 0.25

and 0.41; but when the incidence is 0.05, the derived heritability of liability is 0.10, 0.27 and 0.45 (McGuirk, 1989).

Inclusion of fixed effects in the model

Approximation of the heritability for liability from the heritability of the binary trait using

$$h_L^2 = \frac{p(1-p)}{z^2} h_{01}^2$$

is appropriate if all animals belong to the same group. However, if the incidence of the binary trait depends on the group of animals being measured, such as animals of one herd measured in different years, then derivation of the heritability for liability by transformation of the heritability for the binary trait is inappropriate, as the transformation is a function of the incidence, which is different for each group of animals.

As discussed in Chapter 11, mixed model procedures, such as REML, can simultaneously estimate fixed and random effects and provide an estimate of the heritability for a normally distributed trait. Given the mixed model equation

$$y = Xb + Zu + e$$

then a change in the level of a fixed effect, such as between boars and gilts, will directly correspond to a change in the measured trait, such as growth rate.

In the analysis of a binary trait, with the assumption of a threshold model, then differences between levels of fixed effects will correspond to changes in the liability. Therefore, a function to link changes in the liability to changes in the binary trait needs to be incorporated in the mixed model equations. One such link function is the logit function,

$$y = \frac{e^\theta}{1+e^\theta} \quad \text{with} \quad \theta = Xb + Zu + e$$

where θ represents the liability and y represents the probability that the animal belongs to category one of the binary trait.

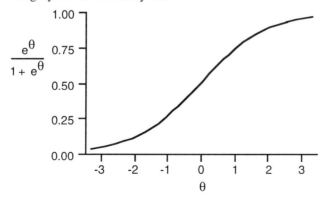

Fig. 14.4. The curve of the logit function

The curve of the logit function, illustrated in Fig. 14.4, indicates that extreme values of the liability correspond to probabilities of zero and unity of an individual being in category one. The liability, θ, can be expressed in terms of y,

since
$$\theta = \log \frac{y}{1-y}$$

For example, if the effect of two treatments on the resistance to a disease is modelled using a threshold model, and estimates of the treatment effect, $\hat{\theta}$, are 2.5 and 3.0 on the liability scale, then the probability of not contracting the disease is 0.92 and 0.95, respectively. However, if estimates of the treatment effect are 1.0 and 1.5 on the liability scale, then the probability of not contracting the disease is 0.73 and 0.82. Therefore, there is not a linear relationship between the estimated treatment effect and disease resistance.

There are several procedures for the analysis of binary data, using a threshold model, to estimate variance components (e.g. Gilmour *et al.*, 1985; Foulley *et al.*, 1987). The relationship between a linear model for the analysis of a normally distributed trait and a threshold model, incorporating a link function, for the analysis of a binary trait is illustrated with the Schall (1991) method.

Firstly, the model for a normally distributed trait is

$$y = Xb + Zu + e$$

and the mixed model equations, as outlined in Chapter 11, are

$$\begin{bmatrix} X'R^{-1}X & X'R^{-1}Z \\ Z'R^{-1}X & Z'R^{-1}Z + G^{-1} \end{bmatrix} \begin{bmatrix} b \\ u \end{bmatrix} = \begin{bmatrix} X'R^{-1}y \\ Z'R^{-1}y \end{bmatrix}$$

where $\text{var}[e] = R = I\sigma_e^2$ and $\text{var}[u] = A\sigma_u^2 = G$.

Variance component estimates are determined from functions of the estimated random effects, $\hat{u}$ and $\hat{e}$. When the random effects, u, are uncorrelated, one pair of functions is

$$\hat{\sigma}_u^2 = \frac{\hat{u}'\hat{u}}{q - \text{tr}(C^{22})/\sigma_u^2}$$

and

$$\hat{\sigma}_e^2 = \frac{e'e}{N - r(X) - q + \text{tr}(C^{22})/\sigma_u^2} \sigma_e^2$$

where q is the number of random effects.

However, for a binary trait, y, the model for the liability, θ, is

$$\theta = Xb + Zu + e \quad \text{and} \quad y = \frac{e^\theta}{1 + e^\theta}$$

The liability is linked to the binary trait by the link function

$$\theta = \log \frac{y}{1-y}$$

Rather than calculating the liability directly, the link function is replaced by a first order Taylor series, evaluated at the predicted value of each observation, $\hat{y}_i$:

$$\theta_i = \log\left(\frac{\hat{y}_i}{1-\hat{y}_i}\right) + \frac{(y_i - \hat{y}_i)}{\hat{y}_i(1-\hat{y}_i)}$$

The mixed model equations for the liability are

$$\begin{bmatrix} X'R^{-1}X & X'R^{-1}Z \\ Z'R^{-1}X & Z'R^{-1}Z+G^{-1} \end{bmatrix} \begin{bmatrix} b_L \\ u_L \end{bmatrix} = \begin{bmatrix} X'R^{-1}\theta \\ Z'R^{-1}\theta \end{bmatrix}$$

where the ith diagonal term of R^{-1} is $\hat{y}_i(1-\hat{y}_i)\dfrac{1}{\sigma_{e_L}^2}$

and

$$\text{var}[u_L] = A\sigma_{u_L}^2 = G$$

with the residual variance on the liability scale equal to $\sigma_{e_L}^2$.

Variance components for the liability are estimated from functions of the estimated random effects:

$$\hat{\sigma}_{u_L}^2 = \frac{\hat{u}_L'\hat{u}_L}{q - \text{tr}(C^{22})/\sigma_{u_L}^2}$$

and

$$\hat{\sigma}_{e_L}^2 = \frac{\hat{e}_L'R^{-1}\hat{e}_L}{N - r(X) - q + \text{tr}(C^{22})/\sigma_{u_L}^2}\sigma_{e_L}^2$$

Note that the form of the formulae for the estimated variance components, $\hat{\sigma}_{u_L}^2$ and $\hat{\sigma}_{e_L}^2$, on the liability scale is same as for the estimated variance components, $\hat{\sigma}_u^2$ and $\hat{\sigma}_e^2$, for a normally distributed trait.

The iterative process is repeated until convergence, with the predicted value of θ, equal to $X\hat{b}_L + Z\hat{u}_L$, transformed, using the link function, to obtain the new predicted value of y, equal to $\dfrac{e^{\hat{\theta}}}{1+e^{\hat{\theta}}}$, and the updated estimate of R^{-1}.

Linear and Non-Linear Models

Incorporation of the link function in the mixed model equations to estimate variance components and predict breeding values is termed a non-linear model, due to the non-linear relationship between the liability and the probability of an individual belonging to a particular category of the binary trait. The complexity of the non-linear model increases computational time, such that the efficiencies of linear and non-linear models for the prediction of breeding values and responses to selection in a binary trait should be determined with different values for the heritability of liability and incidence of the binary trait.

One method of comparing the efficiencies of the two models is to use simulated data, with the normally distributed trait, the liability, generated with a given heritability, to which a threshold is applied to achieve a binary trait. For example, half-sib data for the liability, with a heritability of h_L^2, can be simulated from

$$\theta_{ij} = 0.5 s_i + e_{ij}$$

where θ_{ij} is the liability of the jth progeny of the ith sire, with the sire genetic and environmental effects, s_i and e_{ij}, being random numbers with normal distributions and variances of h_L^2 and $1 - 0.25 h_L^2$, respectively.

When no fixed effects are included in the data simulation and for a constant or variable number of progeny per sire, sire breeding values for the binary trait have similar ranking when predicted with the linear and non-linear models, independent of the heritability of liability or the incidence of the binary trait (Meijering and Gianola, 1985).

In the more realistic situation, when fixed effects are included in the model with a variable number of progeny per sire, then the ranking of the true and predicted sire breeding values is more similar when breeding values are predicted with the non-linear model than with the linear model. The advantage of the non-linear model over the linear model increases as both the heritability of liability and the incidence of the binary trait decrease. Therefore, breeding values of traits which have both low heritabilities and low incidences, as in reproductive traits like twinning in cattle, should be estimated with a non-linear model rather than with a linear model.

References

Brascamp, E.W. (1984) Selection indices with constraints. *Animal Breeding Abstracts* 52, 645–654.

Brascamp, E.W., Smith, C. and Guy, D.R. (1985) Derivation of economic weights from profit equations. *Animal Production* 40, 175–179.

Brisbane, J.R. and Gibson, J.P. (1995) Balancing selection response and rate of inbreeding by including genetic relationships in selection decisions. *Theoretical and Applied Genetics* 91, 421–431.

Bulmer, M.G. (1971) The effect of selection on genetic variability. *The American Naturalist* 105, 201–211.

Burrow, H.M. (1993) The effects of inbreeding in beef cattle. *Animal Breeding Abstracts* 61, 737–751.

Cameron, N.D. and Curran, M.K. (1994) Selection for components of efficient lean growth rate in pigs. 4. Genetic and phenotypic parameter estimates and correlated responses in performance test traits with *ad-libitum* feeding. *Animal Production* 59, 281–291.

Cameron, N.D. and Curran, M.K. (1995) Genotype with feeding regime interaction in pigs divergently selected for components of efficient lean growth rate. *Animal Science* 61, 123–132.

Cameron, N.D., Curran, M.K. and Kerr, J.C. (1994) Selection for components of efficient lean growth rate in pigs. 3. Responses to selection with a restricted feeding regime. *Animal Production* 59, 271–279.

Cunningham, E.P. (1972) Theory and application of statistical selection methods. *XIV British Poultry Breeders Round Table*.

Cunningham, E.P. (1975) Multi-stage index selection. *Theoretical and Applied Genetics* 46, 55–61.

Cunningham, E.P., Moen, R.A. and Gjedrem, T. (1970) Restriction of selection indexes. *Biometrics* 26, 67–74.

Ducos, A., Bidanel, J.P., Ducrocq, V., Boichard, D. and Groeneveld, E. (1993) Multivariate restricted maximum likelihood estimation of genetic parameters for growth, carcass and meat quality traits in French Large White and French Landrace pigs. *Genetics, Selection, Evolution* 25, 475–493.

Eikelenboom, G. and Minkema, D. (1974) Prediction of pale, soft and exudative muscle with a non-lethal test for the halothane induced porcine malignant hyperthermia syndrome. *Netherlands Journal of Veterinary Science* 99, 421–426.

Falconer, D.S. and MacKay, T.F.C. (1996) *Introduction to Quantitative Genetics*. 4th Edition. Longman.

Fernando, R.L. and Grossman, M. (1989) Marker assisted selection using best linear unbiased prediction. *Genetics, Selection, Evolution* 21, 467–477.

Foulley, J.L. (1992) Prediction of selection response for threshold dichotomous traits. *Genetics* 132, 1187–1194.

Foulley, J.L., Im, S., Gianola, D. and Hoschele, I. (1987) Emphirical Bayes estimation of parameters for n polygenic binary traits. *Genetique, Selection, Evolution* 19, 197–224.

Gianola, D. (1982) Theory and analysis of threshold characters. *Journal of Animal Science* 54, 1079–1096.

Gilmour, A.R., Anderson, R.D. and Rae, A.L. (1985) The analysis of binomial data by a generalized linear mixed model. *Biometrics* 72, 593–599.

Goddard, M.E. (1992) A mixed model for analyses of data on multiple genetic markers. *Theoretical and Applied Genetics* 83, 878–886.

Graser, H.-U., Smith, S.P. and Tier, B. (1987) A derivative-free approach for estimating variance components in animal models by restricted maximum likelihood. *Journal of Animal Science* 64, 1362–1370.

Groeneveld, E., Kovac, M. and Wang, T. (1990) PEST, a general purpose BLUP package for multivariate prediction and estimation. *Proceedings of the 4th World Congress on Genetics Applied to Livestock Production* 13, 488–491.

Grundy, B., Caballero, A., Santiago, E. and Hill, W.G. (1994) A note on using biased parameter values and non-random mating to reduce rates of inbreeding in selection programmes. *Animal Production* 59, 465–468.

Haley, C.S. and Lee, G.J. (1992) Genetic factors contributing to variation in litter size in British Large White gilts. *Livestock Production Science* 30, 99–113.

Hanset, R. and Michaux, C. (1985) On the genetic determinism of muscular hypertrophy in the Belgian White and Blue cattle breed. 1. Experimental data. *Genetique, Selection, Evolution* 17, 359–368.

Harville, D.A. (1977) Maximum Likelihood approaches to variance component estimation and to related problems. *Journal of the American Statistical Association* 72, 320–340.

Hayes, J.F. and Hill, W.G. (1980) A reparameterization of a genetic selection index to locate its sampling properties. *Biometrics* 36, 237–248.

Hayes, J.F. and Hill, W.G. (1981) Modification of estimates of parameters in the construction of genetic selection indices ('bending'). *Biometrics* 37, 483–493.

Hazel, L.N. (1943) The genetic basis of constructing selection indexes. *Genetics* 28, 476–490.

Henderson, C.R. (1953) Estimation of variance and covariance components. *Biometrics* 9, 226–252.

Henderson, C.R. (1973) Sire evaluation and genetic trends. *Proceedings of the Animal Breeding and Genetics Symposium in Honour of Dr. J.L. Lush*, ASAS and ASDA, pp. 10–41.

Henderson, C.R. (1975) Best linear unbiased estimation and prediction under a selection model. *Biometrics* 31, 423–447.

Henderson, C.R. (1976) A simple method for computing the inverse of a numerator relationship matrix used in prediction of breeding values. *Biometrics* 32, 69–83.

Henderson, C.R. and Quaas, R.L. (1976) Multiple trait evaluation using relatives' records. *Journal of Animal Science* 43, 1188–1197.

Hill, W.G. (1979) A note on the effective population size with overlapping generations. *Genetics* 92, 317–322.

Hill, W.G. and Thompson, R. (1977) Design of experiments to estimate offspring-parent regression using selected parents. *Animal Production* 24, 163–168.

Hill, W.G. and Thompson, R. (1978) Probabilities of non-positive definite between-group or genetic covariance matrices. *Biometrics* 34, 429–439.

Hill, W.G. and Webb, A.J. (1982) Genetics of reproduction in the pig. In: Cole, D.J.A. and Foxcroft, G.R. (eds), *Control of Pig Reproduction*. pp. 541–564, Butterworth.

Hill, W.G., Caballero, A. and Dempfle, L. (1996) Prediction of response to selection within families. *Genetics, Selection, Evolution* 28, 379–383.

Hovenier, R., Kanis, E., van Asseldonk, Th. and Westerink, N.G. (1992) Genetic parameters of pig meat quality traits in a halothane negative population. *Livestock Production Science* 32, 309–321.

Janss, L.L.G., van Arendonk, J.A.M. and Brascamp, E.W. (1994) Identification of a single gene affecting intramuscular fat in Meishan crossbreds using Gibbs sampling. *Proceedings of the 5th World Congress on Genetics Applied to Livestock Production* 18, 361–364.

Kempthorne, O. and Nordskog, A.W. (1959) Restricted selection indices. *Biometrics* 15, 10–19.

Kerr, J.C. and Cameron, N.D. (1996) Genetic and phenotypic relationships between performance test and reproduction traits in Large White pigs. *Animal Science* 62, 531–540.

Lander, E.S. and Botstein, D. (1989) Mapping Mendelian factors underlying quantitative traits using RFLP linkage maps. *Genetics* 121, 185–199.

Le Roy, P., Naveau, J., Elsen, J.M. and Sellier, P. (1990) Evidence for a new major gene influencing meat quality in pigs. *Genetical Research* 55, 33–40.

MacLennan, D.H., Duff, C., Zorzato, F., Fujii, J., Phillips, M., Korneluk, R.G., Frodis, W., Britt, B.A. and Worton, R.G. (1990) Ryanodine receptor gene is a candidate for predisposition to malignant hyperthermia. *Nature* 343, 559–561.

McGuirk, B.J. (1989) The estimation of genetic parameters for all-or-none and categorical traits. In: Hill, W.G. and MacKay, T.F.C. (eds), *Evolution and Animal Breeding : Reviews on Molecular and Quantitative Approaches in Honour of Alan Robertson*. pp. 175–180, CAB International.

Meijering, A. (1986) Revenues from sire selection for calving traits in Dutch dairy cattle. *Journal of Animal Breeding and Genetics* 103, 358–375.

Meijering, A. and Gianola, D. (1985) Linear versus nonlinear methods of sire evaluation for categorical traits: a simulation study. *Genetique, Selection, Evolution* 17, 115–132.

Meuwissen, T.H.E. and Goddard, M.E. (1996) The use of marker haplotypes in animal breeding schemes. *Genetics, Selection, Evolution* 28, 161–176.

Meuwissen, T.H.E. and van Arendonk, J.A.M. (1992) Potential improvements in rate of genetic gain from marker-assisted selection in dairy cattle breeding schemes. *Journal of Dairy Science* 75, 1651–1659.

Meuwissen, T.H.E. and Woolliams, J.A. (1994) Response versus risk in breeding schemes. *Proceedings of the 5th World Congress on Genetics Applied to Livestock Production* 18, 236–243.

Meyer, K. (1985) Maximum likelihood estimation of variance components for a multivariate mixed model with equal design matrices. *Biometrics* 41, 153–165.

Meyer, K. (1989) Restricted maximum likelihood to estimate variance components for animal models with several random effects using a derivative-free algorithm. *Genetics, Selection, Evolution* 21, 317–340.

Meyer, K. (1992) Variance components due to direct and maternal effects for growth traits of Australian beef cattle. *Livestock Production Science* 31, 179–204.

Meyer, K. and Thompson, R. (1984) Bias in variance and covariance component estimators due to selection on a correlated trait. *Zeitschrift für Tierzuchtung und Zuchtungsbiologie* 101, 33–50.

Miglior, F., Burnside, E.B. and Kennedy, B.W. (1995) Production traits of Holstein cattle–estimation of nonadditive genetic variance components and inbreeding depression. *Journal of Dairy Science* 78, 1174–1180.

Mrode, R.A. (1996) *Linear Models for the Prediction of Animal Breeding Values.* CAB International.

Nicholas, F.W. and Smith, C. (1983) Increased rates of genetic change in dairy cattle by embryo transfer and splitting. *Animal Production* 36, 341–353.

Patterson, H.D. and Thompson, R. (1971) Recovery of inter-block information when block sizes are unequal. *Biometrika* 58, 545–554.

Piper, L.R., Bindon, B.M. and Davies, G.H. (1985) The single gene inheritance of the high litter size of the Booroola Merino. In: Land, R.B and Robinson, D.W. (eds), *Genetics of Reproduction in Sheep.* pp. 115–125, Butterworth.

Quaas, R.L. (1976) Computing the diagonal elements and inverse of a large numerator relationship matrix. *Biometrics* 32, 949–953.

Quaas, R.L. and Pollak, E.J. (1980) Mixed model methodology for farm and ranch beef cattle testing programs. *Journal of Animal Science* 51, 1277–1287.

Rao, C.R. (1973) *Linear Statistical Inference and its Applications.* 2nd edition. John Wiley & Sons.

Robertson, A. (1959a) Experimental design in the evaluation of genetic parameters. *Biometrics* 15, 219–226.

Robertson, A. (1959b) The sampling variance of the genetic correlation coefficient. *Biometrics* 15, 469–485.

Robertson, A. (1977) The effect of selection on the estimation of genetic parameters. *Zeitschrift für Tierzuchtung und Zuchtungsbiologie* 94, 131–135.

Robertson, A. and Lerner, I.M. (1949) The heritability of all-or-none traits: Viability of poultry. *Genetics* 34, 395–411.

Rothschild, M., Jacobson, C., Vaske, D., Tuggle, C., Wang, L.Z., Short, T., Eckardt, G., Sasaki, S., Vincent, A., McLaren, D., Southwood, O., van der Steen, H., Mileham, A. and Plastow, G. (1996) The estrogen-receptor locus is associated with a major gene influencing litter size in pigs. *Proceedings of the National Academy of Sciences of the United States of America* 93, 201–205.

Sales, J. and Hill, W.G. (1976a) Effect of sampling errors on efficiency of selection indices. 1. Use of information from relatives for single trait improvement. *Animal Production* 22, 1–17.

Sales, J. and Hill, W.G. (1976b) Effect of sampling errors on efficiency of selection indices. 2. Use of information on associated traits for improvement of a single important trait. *Animal Production* 23, 1–14.

Schaeffer, L.R., Wilton, J.W. and Thompson, R. (1978) Simultaneous estimation of variance and covariance components from multitrait mixed model equations. *Biometrics* 34, 199–208.

Schall, R. (1991) Estimation in generalized linear models with random effects. *Biometrika* 78, 719–727.

Searle, S.R. (1971) *Linear Models*. John Wiley & Sons.

Smith, C. (1983) Effects of changes in economic weights on the efficiency of index selection. *Journal of Animal Science* 56, 1057–1064.

Smith, C., James, J.W. and Brascamp, E.W. (1986) On the derivation of economic weights in livestock improvement. *Animal Production* 43, 545–551.

Snedecor, G.W. and Cochran, W.G. (1989) *Statistical Methods*. 8th edition. The Iowa State University Press.

Swalve, H.H. (1995) The effect of test day models on the estimation of genetic parameters and breeding values for dairy yield traits. *Journal of Dairy Science* 78, 929–938.

Thompson, R. (1976) The estimation of maternal genetic variances. *Biometrics* 32, 903–917.

Thompson, R. (1979) Sire evaluation. *Biometrics* 35, 339–353.

Thompson, R., Wray, N.R. and Crump, R.E. (1994) Calculation of prediction error variances using sparse matrix methods. *Journal of Animal Breeding and Genetics* 111, 102–109.

Thompson, R., Crump, R.E., Juga, J. and Visscher, P.M. (1995) Estimating variances and covariances for bivariate animal models using scaling and transformation. *Genetics, Selection, Evolution* 27, 33–42.

Toro, M. and Perez–Encisco, M. (1990) Optimization of selection response under restricted inbreeding. *Genetics, Selection, Evolution* 22, 93–107.

van Arendonk, J.A.M., Tier, B. and Kinghorn, B.P. (1994) Use of multiple genetic markers in prediction of breeding values. *Genetics* 137, 319–329.

van der Beek, S. and van Arendonk, J.A.M. (1996) Marker-assisted selection in an outbred poultry breeding nucleus. *Animal Science* 62, 171–180.

Weller, J.I. (1994) *Economic Aspects of Animal Breeding.* Chapman and Hall.

Weller, J.I., Kashi, Y. and Soller, M. (1990) Power of daughter and granddaughter designs for determining linkage between marker loci and quantitative trait loci in dairy cattle. *Journal of Dairy Science* 73, 2525–2537.

Westell, R.A., Quaas, R.L. and Van Vleck, L.D. (1988) Genetic groups in an animal model. *Journal of Animal Science* 71, 1310–1318.

Woolliams, J.A. (1989) Modifications to MOET nucleus breeding schemes to improve rates of genetic progress and decrease rates of inbreeding in dairy cattle. *Animal Production* 49, 1–14.

Woolliams, J.A. and Pong-Wong, R. (1995) Short– versus long–term responses in breeding schemes. *Proceedings of the 46th Annual Meeting of the European Association for Animal Production* 1, 35.

Woolliams, J.A. and Smith, C. (1988) The value of indicator traits in the genetic improvement of dairy cattle. *Animal Production* 46, 333–345.

Woolliams, J.A., Wray, N.R. and Thompson, R. (1993) Prediction of long–term contributions and inbreeding in populations undergoing mass selection. *Genetical Research, Cambridge* 62, 231–242.

Wray, N.R., Woolliams, J.A. and Thompson, R. (1994) Prediction of rates of inbreeding in populations undergoing index selection. *Theoretical and Applied Genetics* 87, 878–892.

Appendix

Matrix Algebra

Matrix notation can be used to represent equations in an uncomplicated manner, such that an understanding of the properties of an equation can be obtained without getting distracted by detailed algebra. Secondly, obtaining solutions to equations is routinely performed through the use of matrices.

A very simple example of the use of matrices can be taken from school mathematics. If one person buys nine oranges and four apples and pays 158p and a second person buys five oranges and six apples and pays 118p, then what are the costs of oranges and apples? The standard approach is to set up two equations describing the purchases

$$9 \text{ oranges} + 4 \text{ apples} = 158p$$
$$5 \text{ oranges} + 6 \text{ apples} = 118p$$

The equations are solved by eliminating one of the variables from the equations and then solving for the other variable. To eliminate apples from the equations, the first equation is multiplied by six and the second is multiplied by four; then the first equation is subtracted from the second equation to provide

$$34 \text{ oranges} = 476$$

such that oranges cost 14p. The cost of oranges is substituted in either equation, to enable the cost of apples to be determined, which is 8p.

Before illustrating how matrix algebra can be used to solve the problem, some definitions are required.

Definition of a matrix

A matrix consists of R rows and C columns of numbers. For example, if matrix A was equal to $\begin{bmatrix} 1 & 2 & 3 \\ 4 & 5 & 6 \end{bmatrix}$, then A would be a 2×3 matrix, as the number of rows proceeds the number of columns. The element of matrix A in row i and column j is denoted as A_{ij}. A vector is just a matrix with one column. To differentiate between matrices, vectors and scalars or numbers, in the Appendix, matrices and vectors are in bold.

Matrix multiplication

If A and B are two matrices, then the product of A and B is C = AB. For matrix multiplication, the number of columns of A must equal the number of rows of B. The element of matrix C in the ith row and j^{th} column, C_{ij}, is the product of the elements in row i of matrix A multiplied by the elements in column j of matrix B.

$$\text{If } A = \begin{bmatrix} 1 & 2 & 3 \\ 4 & 5 & 6 \end{bmatrix} \text{ and } B = \begin{bmatrix} 2 & 4 & 6 & 8 \\ 3 & 5 & 7 & 9 \\ 2 & 3 & 6 & 7 \end{bmatrix}, \text{ then } C = \begin{bmatrix} 14 & 23 & 38 & 47 \\ 35 & 59 & 95 & 119 \end{bmatrix}.$$

For example, the third element in the second row of matrix C, equal to 95, is calculated as $(4 \times 6) + (5 \times 7) + (6 \times 6)$. The dimensions of matrix C are equal to the rows of matrix A and the columns of matrix B. Even when A and B are square matrices, the product AB does not always equal the product BA.

The inverse of a matrix

Only square matrices can have an inverse, since the product of matrix A and its inverse is the identity matrix, such that

$$AA^{-1} = A^{-1}A = I$$

where the identity matrix, I, is a square matrix, with ones on the diagonal and the off-diagonal elements are equal to zero.

The inverse of the 2×2 matrix, $\begin{bmatrix} a & b \\ c & d \end{bmatrix}$, is

$$\frac{1}{ad - bc} \begin{bmatrix} d & -b \\ -c & a \end{bmatrix}$$

where $(ad - bc)$ is the determinant of the matrix. If the determinant is zero, then the matrix has no inverse. Calculation of the inverse of matrices larger than 2×2 matrices is generally performed using computer programs.

Returning to the apple and oranges problem, the equations can be expressed in the form $A x = y$, where x is the vector representing the cost of oranges and apples, $x = \begin{bmatrix} \text{oranges} \\ \text{apples} \end{bmatrix}$, y is the price of the two purchases, $y = \begin{bmatrix} 158 \\ 118 \end{bmatrix}$, and A is the matrix of coefficients for the two purchases, with each row representing one purchase, with $A = \begin{bmatrix} 9 & 4 \\ 5 & 6 \end{bmatrix}$, such that

$$\begin{bmatrix} 9 & 4 \\ 5 & 6 \end{bmatrix} \begin{bmatrix} \text{oranges} \\ \text{apples} \end{bmatrix} = \begin{bmatrix} 158 \\ 118 \end{bmatrix}$$

The inverse of A is

$$\frac{1}{(9 \times 6) - (4 \times 5)} \begin{bmatrix} 6 & -4 \\ -5 & 9 \end{bmatrix} = \begin{bmatrix} 0.176 & -0.118 \\ -0.147 & 0.265 \end{bmatrix}$$

Multiplying both sides of the equation by the inverse of the A matrix, we obtain

$$\begin{bmatrix} \text{oranges} \\ \text{apples} \end{bmatrix} = \begin{bmatrix} 14 \\ 8 \end{bmatrix}$$

Obviously, the example is simple, but it clearly illustrates that solutions to equations can be determined through the use of matrices, particularly when there are a large number of equations and variables.

Four other matrix terms need to be defined, which will be required for selection indices and the prediction of breeding values.

Multiplying a matrix by a constant

If matrix A is multiplied by a constant c, then all the elements of the matrix A are multiplied by c.

Addition and subtraction of matrices

Matrices can only be added or subtracted if the dimensions of the two matrices are equal. The element of (A + B) in the ith row and j^{th} column is $A_{ij} + B_{ij}$, such that A + B is equivalent to B + A. Similarly, the element of (A − B) in the ith row and j^{th} column is $A_{ij} - B_{ij}$.

Transpose of a matrix

The transpose of matrix A, with R rows and C columns, is obtained by replacing the rows by the columns, such that the transpose has C rows and R columns. The element in the ith row and j^{th} column of the transpose of A is equal to the element in the jth row and i^{th} column of A. The transpose of matrix A is denoted by A'.

The matrix A is symmetric if A = A'.

Variance–covariance matrices

If $X_1, X_2, \ldots, X_n$ are variables, with variances $\text{var}(X_1), \text{var}(X_2), \ldots, \text{var}(X_n)$ and covariances $\text{cov}(X_1, X_2), \text{cov}(X_1, X_3), \ldots, \text{cov}(X_{n-1}, X_n)$, then the variance–covariance matrix of $X_1, X_2, \ldots, X_n$ is a square matrix, with the variances on the diagonal and the covariances on the off-diagonal:

$$\begin{bmatrix} \text{var}(X_1) & \text{cov}(X_1,X_2) & \cdots & \text{cov}(X_1,X_n) \\ \text{cov}(X_1,X_2) & \text{var}(X_2) & \cdots & \text{cov}(X_2,X_n) \\ \vdots & \vdots & \vdots & \vdots \\ \text{cov}(X_1,X_n) & \text{cov}(X_2,X_n) & \cdots & \text{var}(X_n) \end{bmatrix}$$

The element in the ith row and j^{th} column of the variance–covariance matrix is equal to the covariance between X_i and X_j.

Example

The variance of a linear combination variables can be represented easily using matrix algebra. If $X_1, X_2, \ldots, X_n$ are variables with variance–covariance matrix A, and $c_1, c_2, \ldots, c_n$ are constants represented by the vector c, then the variance of the linear combination

$$\text{var}(c_1 X_1 + c_2 X_2 + \ldots + c_n X_n) = c'Ac$$

For example, the variance of the linear combination is the sum of the elements of the symmetric matrix

$$\begin{bmatrix} c_1^2 \text{var}(X_1) & c_1 c_2 \text{cov}(X_1,X_2) & c_1 c_3 \text{cov}(X_1,X_3) & c_1 c_4 \text{cov}(X_1,X_4) \\ & c_2^2 \text{var}(X_2) & c_2 c_3 \text{cov}(X_2,X_3) & c_2 c_4 \text{cov}(X_2,X_4) \\ & & c_3^2 \text{var}(X_3) & c_3 c_4 \text{cov}(X_3,X_4) \\ & & & c_4^2 \text{var}(X_4) \end{bmatrix}$$

An example of a matrix representation of the variance of the linear combination is the variance of a sib mean, described in Chapter 5. If there are three sibs, then the phenotypic variance–covariance matrix of the sibs is:

$$\begin{bmatrix} \sigma_P^2 & t\sigma_P^2 & t\sigma_P^2 \\ t\sigma_P^2 & \sigma_P^2 & t\sigma_P^2 \\ t\sigma_P^2 & t\sigma_P^2 & \sigma_P^2 \end{bmatrix}$$

The variance of the sib mean, $\text{var}\left(\frac{1}{3}X_1 + \frac{1}{3}X_2 + \frac{1}{3}X_3\right)$, is the sum of the elements of the matrix

$$\begin{bmatrix} \frac{1}{9}\sigma_P^2 & \frac{1}{9}t\sigma_P^2 & \frac{1}{9}t\sigma_P^2 \\ \frac{1}{9}t\sigma_P^2 & \frac{1}{9}\sigma_P^2 & \frac{1}{9}t\sigma_P^2 \\ \frac{1}{9}t\sigma_P^2 & \frac{1}{9}t\sigma_P^2 & \frac{1}{9}\sigma_P^2 \end{bmatrix}$$

which is

$$\frac{1}{n^2}\left[n\sigma_P^2 + n(n-1)t\sigma_P^2\right] = \left[\frac{1+(n-1)t}{n}\right]\sigma_P^2.$$

Normal Distribution Tables

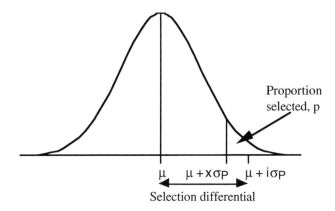

Selection differential

If the proportion of animals selected, p, have measurements greater than the population mean by x standard deviations, then the mean of the selected animals will be greater than the population mean by i standard deviations.

If the proportion of animals selected, p, is greater than 0.5, then use the tabulated values for $1 - p$, multiply x by -1 and multiply i by $(1 - p)/p$.

p	x	i	p	x	i	p	x	i
1	2.33	2.66	21	0.81	1.37	41	0.23	0.95
2	2.05	2.42	22	0.77	1.35	42	0.20	0.93
3	1.88	2.27	23	0.74	1.32	43	0.18	0.91
4	1.75	2.15	24	0.71	1.30	44	0.15	0.90
5	1.64	2.06	25	0.67	1.27	45	0.13	0.88
6	1.56	1.98	26	0.64	1.25	46	0.10	0.86
7	1.48	1.92	27	0.61	1.22	47	0.08	0.85
8	1.40	1.86	28	0.58	1.20	48	0.05	0.83
9	1.34	1.80	29	0.55	1.18	49	0.02	0.81
10	1.28	1.76	30	0.52	1.16	50	0.00	0.80
11	1.23	1.71	31	0.50	1.14			
12	1.18	1.67	32	0.47	1.12			
13	1.13	1.63	33	0.44	1.10			
14	1.08	1.59	34	0.41	1.08			
15	1.04	1.55	35	0.38	1.06			
16	0.99	1.52	36	0.36	1.04			
17	0.95	1.49	37	0.33	1.02			
18	0.92	1.46	38	0.30	1.00			
19	0.88	1.43	39	0.28	0.98			
20	0.84	1.40	40	0.25	0.97			

Questions

Q1. Growth rates of pigs were recorded, with an equal number of animals in each litter. A balanced ANOVA table was calculated from the data:

Source of variation	DF	Mean squares
Between-sires	16	340.75
Between-dams within-sires	51	175.75
Between-progeny within-dams	136	92.5

(a) How many sires, dams per sire and progeny per dam were there?
(b) What are the expectations of the mean squares?
(c) Calculate the variance components and the phenotypic variance for growth rate.

Q2. Using the data in Question 1, calculate:
(a) the variance of a sire family mean for growth rate;
(b) the variance of a dam family mean for growth rate;
(c) the half-sib and full-sib correlations for growth rate.
(d) Determine whether the half-sib correlation is different from zero.

Q3. Using the data in Question 1, calculate:
(a) the half-sib heritability and its standard error;
(b) the maternal half-sib heritability and its standard error;
(c) the full-sib heritability and its standard error;
(d) the maternal effect, c^2, and its standard error.

Q4. Calculate the standard error of a heritability estimated from half-sib data, when the true heritability is 0.25, the phenotypic variance is 144 and the number of sires and progeny per sire are 20 and 30, respectively.

Q5. Litter size measurements for dams and the average of their daughters, with 30 daughters per dam are as follows:

Pair	1	2	3	4	5	6	7	8	9	10
Dam	11	9	13	10	9	8	10	11	10	13
Daughter average	10.0	9.7	10.2	9.9	9.8	9.8	10.0	10.1	9.8	10.4

Calculate:
(a) the regression equation for average daughter litter size on dam litter size;
(b) the correlation between the dam and average daughter litter sizes;
(c) the standard error of the regression coefficient and the 0.95 confidence interval of the correlation coefficient;
(d) the heritability of litter size using the regression and correlation coefficients.

Q6. A balanced analysis of covariance was calculated using data on growth rate and food conversion ratio, FCR, from animals in Question 1:

Source of variation	DF	Growth rate mean squares	Mean cross-products	FCR mean squares
Between-sires	16	340.75	−431.5	1219.0
Between-dams	51	175.75	−197.5	559.0
Between-progeny	136	92.5	−121.0	325.0

(a) Calculate the covariance components.
(b) Calculate the phenotypic, genetic and environmental correlations, using half-sib heritability estimates.
(c) Calculate the standard error of the genetic correlation.

Q7. The litter weights of daughter's litters and dam's litters were measured in two experiments:

Experiment 1: all litters were standardised at birth to eight progeny, by removing progeny in excess of eight or by fostering from other litters.
Experiment 2: all mothers reared all their progeny.

The regression coefficients of daughter's litter weight on their dam's litter weight were 0.049 and −0.028 in the two experiments and the repeatability, or correlation, of litter weights for full-sibs were 0.054 and 0.081, respectively.
Suggest reasons, other than sampling variation, which account for the differences between the heritability estimates.

Q8. In a study of fertility in pigs, the same number of litter sizes was measured on each of 156 unrelated sows. The ANOVA table was as follows:

Source of variation	DF	Mean squares
Between-dams	155	25.56
Within-dams	1404	3.23

(a) Calculate the repeatability of litter size.

(b) In the design of a breeding programme to improve fertility, a high heritability is desirable. By how much would the heritability of fertility be increased if fertility was defined as the mean of two, three or four litters?

Q9. Calculate the expected response to selection for increased body weight, given a heritability of 0.26, a phenotypic variance of 10.6 and that the proportions of individuals selected are 0.25, 0.50 and 0.75.

Calculate the expected responses if five measurements are made on each individual, and assume that the repeatability is equal to the heritability.

Q10. Prove that
$$R = ib_{AS}\sigma_S = ir_{AS}\sigma_A = i\sigma_{\hat{A}}$$
where A is the animal's true breeding value, $\hat{A}$ is the animal's predicted breeding value and S is the selection criterion.

Q11. The mean values of a trait for animals A and B are 520 and 500 kg, with 5 and 15 measurements, respectively. The heritability of the trait is 0.2, the repeatability is 0.3, the phenotypic variance is 350 kg^2 and the mean value for the population is 420 kg. For animals A and B, calculate:
(a) the predicted breeding values, BV;
(b) the variance of the predicted BV;
(c) the accuracy of the predicted BV;
(d) the prediction error variance;
(e) the 95% confidence interval of the BV.

Q12. Average daily gain in pigs has a half-sib correlation of 0.1 and a full-sib correlation of 0.36. Calculate the regression coefficients, $b_{A\bar{F}}$ and $b_{A\bar{S}}$, for an individual given the following information on relatives:
(a) the mean of five half-sibs, all from different dams, with the individual included in the mean;
(b) the mean of five full-sibs, which includes the individual;
(c) the mean of four full-sibs, which excludes the individual;
(d) the individual's deviation from the mean of five full-sibs, which includes the individual.

Q13. Calculate the rates of response, relative to selection on the individual's measurement only, for each situation in Question 12, ignoring differences in the selection intensity.

Questions

Q14. Which of the two pigs, A or B, would be selected using the regression coefficients from Question 12, parts (b), (c) or (d), or if only the individual's measurement was used?

Pig	Growth rate	Full-sib * mean growth rate	Population mean (g/day)
A	980	920	950
B	970	990	

* Includes the individual's measurement.

Q15. Which bull would be selected, given a heritability for milk yield of 0.25, a phenotypic standard deviation of 1400 kg and a population mean of 8000 kg, when the mean progeny milk yields of bulls A and B are 8300 and 8500 kg, for 50 and 10 progeny, respectively?

Q16. Average daily gain in pigs has a half-sib correlation of 0.10 and a full-sib correlation of 0.36. Calculate selection criterion coefficients and the correlation between the selection criterion and the selection objective, given the following measurements:
(a) the mean of five half-sibs with the individual excluded and the individual's measurement;
(b) the mean of five half-sibs with the individual excluded and the deviation of the individual's measurement from the half-sib mean;
(c) the individual's measurement and the mean of six half-sibs with the individual included.
(d) Repeat parts (a), (b) and (c) with full-sibs.

Q17. Calculate the selection criterion to improve average daily gain, ADG, and food conversion ratio, FCR, when each individual is measured:

	ADG	FCR	σ_P	Economic value
ADG	**0.52**	0.83	1.11	1
FCR	0.71	**0.40**	2.48	5

Heritabilities, in bold, are on the diagonal, the genetic correlation is below the diagonal and the phenotypic correlation is above the diagonal.

Q18. Calculate the correlated responses in average daily gain and food conversion ratio to selection on the criterion in Question 17 and determine the correlation of each trait with the selection criterion.

Q19. Calculate the heritability of the selection criterion in Question 17.

Q20. Using the parameters in Question 17, determine the selection criterion to improve average daily gain only, when measurements are made on average daily gain and food conversion ratio. Calculate the correlated responses in average daily gain and food conversion ratio.

Q21. Average daily gain, ADG, and carcass lean content, LEAN, are to be improved in a pig herd. Ultrasonic backfat depth, BFAT, and average daily gain are measured on each pig. Calculate the correlation between the selection criterion and the selection objective, when one, two, four or six backfat depth measurements are made on each pig. The genetic and phenotypic parameters are as follows:

	ADG	BFAT	LEAN	σ_P	economic value	t_{FS}
ADG	**0.46**	0.06		0.102	980	0.29
BFAT	−0.05	**0.31**		3.48		0.21
LEAN	0.00	−0.60	**0.45**	4.0	60	

Heritabilities, in bold, are on the diagonal, genetic correlations are below the diagonal and phenotypic correlations are above the diagonal.
The repeatability of ultrasonic backfat depth measurements is 0.70.

Q22. As in Question 21, average daily gain, ADG, and carcass lean content, LEAN, are to be improved in a pig herd. Average daily gain and ultrasonic backfat depth, BFAT, are included in a selection criterion using the following parameters, with one ultrasonic backfat depth measurement for each individual:

	ADG	BFAT	ADF	LEAN	σ_P	Economic value
ADG	**0.46**	0.06	0.54		0.102	980
BFAT	−0.05	**0.31**	0.42		3.48	
ADF	0.58	0.54	**0.34**		0.221	
LEAN	0.00	−0.60	−0.45	**0.45**	4.0	60

Calculate the responses for all four traits when average daily food intake, ADF, is included or excluded from the selection criterion, which consists of average daily gain and ultrasonic backfat depth.

Q23. Using the parameters in Question 22, calculate the restricted selection index, which will result in no genetic change in average daily food intake. Compare the correlation between the selection criterion in Question 22 and the selection objective with the correlation for the restricted selection criterion.

Q24. Average gaily gain, ADG, and carcass lean content, LEAN, are to be improved in a pig herd. Measurements for average daily gain and ultrasonic backfat depth, BFAT, are made on the individual and its sibs.

Calculate the selection criterion coefficients and the correlation between the selection objective, when one, two, three or four full-sibs are measured. The individual's measurement is to be included in the full-sib mean.

(Selection criterion = b_{ADG}ind + b_{BFAT}ind + $b_{ADG}\overline{sib}$ + $b_{BFAT}\overline{sib}$.)
Use the parameters given in Question 21.

Q25. Calculate the actual loss and the predicted gain in response by including trait X as well as trait Y in the selection criterion to improve trait Y, when the estimated genetic correlation between traits X and Y is equal to (a) 0.2 and (b) 0.5.

The heritabilities of trait X and Y are 0.4 and 0.1, with the genetic and phenotypic correlations equal to 0.3 and 0.6, respectively.

Q26. Which of the following breeding programmes has the greater annual rate of genetic improvement?

		Programme A	Programme B
Selection of males	r_{IH}	0.30	0.90
Selected proportion	p	0.02	0.10
Generation interval	L	2 years	7 years
Selection of females	r_{IH}	0.60	0.65
Selected proportion	p	0.90	0.90
Generation interval	L	4 years	4 years

Males in programme A were selected on the dam's predicted breeding values. The value of r_{IH} for dams was 0.60.

Q27. Predict the breeding values of the three animals, A, B and C, their prediction error variances and the values of r_{IH}, given repeated measurements on the animal, with one record on each of the animal's progeny and sire, as follows:

Animal	Number of measurements	Average	Number of progeny	Average	Sire's record
A	3	50	30	30	−10
B	4	−20	20	60	40
C	5	70	10	20	30

The average values are expressed as deviations from the population mean. The additive genetic variance, environmental variance and phenotypic variance are equal to 40, 40 and 100, respectively, and all progeny are half-sibs.

Q28. The following measurements were made on five animals:

Animal	Measurements
1	One record = 20
2	Record of sire of animal 2 = 30
	Animal 1 is the dam of animal 2
3	Animal 1 is the dam of animal 3
4	Average of 10 half-sib progeny of animal 4 = 19
5	Average of 10 half-sibs of animal 5 = 19
	Animal 5 is the sire of animal 3

Calculate the breeding values of the five animals, when the heritability is 0.4 and the phenotypic variance is 200, given the above measurements, expressed as deviations from the population mean.

Calculate the prediction error variances and the values of r_{IH} for animals 1 to 5 and the sire of animal 2.

Q29. Write down the mixed model equations to obtain breeding values of the three sires, which have the following records on their progeny in three herds. Assume the sires are unrelated and that the heritability of protein yield is 0.4.

Herd	1	1	1	2	2	2	3	3	3
Sire	A	B	C	A	B	B	C	C	C
Protein	210	160	120	180	115	125	120	100	80

If sire C is the son of sire B, who is a half-sib of sire A, then how would the mixed model equations change?

Q30. Daily milk yield was recorded on several occasions during two months. The two animals, A and B, are half-sibs, the heritability of daily milk yield is 0.30, the repeatability is 0.50 and the phenotypic variance is 40.

	July			August		
Animal A	18	20	20	19		
Animal B		21	22	22	20	19

(a) Determine the elements of the mixed model equations required to predict breeding values of the two animals for daily milk yield.
(b) Discuss the use of the two effects which are estimated for each animal.
(c) Outline the procedure to determine the prediction error variances for the two animals and describe the uses of the prediction error variance.

Answers

For further information, the numbers in [ ] brackets refer to the appropriate pages in the text.

Q1. [13]
(a) Formulae for calculating the degrees of freedom in the ANOVA table are:

Between-sires $(s-1)$ = 16   s = number of sires = 17
Between-dams $s(d-1)$ = 51   d = number of dams per sire = 4
Between-progeny $sd(n-1)$ = 136   n = number of progeny per dam = 3

(b) The expectations of the mean squares are:

Between-sires $\sigma_e^2 + n\sigma_d^2 + nd\sigma_s^2$ = $\sigma_e^2 + 3\sigma_d^2 + 12\sigma_s^2$ = MS_s
Between-dams $\sigma_e^2 + n\sigma_d^2$ = $\sigma_e^2 + 3\sigma_d^2$ = MS_d
Between-progeny σ_e^2 = σ_e^2 = MS_e

(c) $MS_e = 92.5 = \sigma_e^2$

$MS_d = 175.75 = \sigma_e^2 + 3\sigma_d^2$ $\sigma_d^2 = (MS_d - MS_e)/3 = 27.75$

$MS_s = 340.75 = \sigma_e^2 + 3\sigma_d^2 + 12\sigma_s^2$ $\sigma_s^2 = (MS_s - MS_d)/12 = 13.75$

The phenotypic variance, $\sigma_P^2 = \sigma_s^2 + \sigma_d^2 + \sigma_e^2$, equals 134.0.

Q2. [15]
(a) Variance of sire family mean = $\sigma_s^2 + \sigma_d^2/d + \sigma_e^2/nd = 28.40$.
(b) Variance of dam family mean = $\sigma_d^2 + \sigma_e^2/n + \sigma_s^2 = 72.33$.
(c) Half-sib correlation: $t_{HS} = \sigma_s^2/\sigma_P^2$ = 0.103.
   Full-sib correlation: $t_{FS} = (\sigma_s^2 + \sigma_d^2)/\sigma_P^2$ = 0.310.

(d) The variance of the half-sib correlation can be calculated using the formula for the variance of the sire variance component and by treating the estimate of the phenotypic variance as a constant, such that

$$\text{var}(t_s) = \text{var}\left(\frac{\sigma_s^2}{\sigma_P^2}\right) = \frac{\text{var}(\sigma_s^2)}{\sigma_P^4}$$

The formula for the variance of the sire variance component is

$$\frac{2}{n^2 d^2}\left[\frac{MS_s^2}{(s-1)+2} + \frac{MS_d^2}{s(d-1)+2}\right]$$

such that the variance of the sire variance component is 97.7 and the standard error of the half-sib correlation is 0.074.

The 0.95 confidence interval of the half-sib correlation is (−0.04, 0.248), determined from $t_{HS} \pm 1.96 \text{ s.e.}(t_{HS})$, and as zero is contained within the 0.95 confidence interval, then the half-sib correlation is not significantly different from zero at the 0.05 level of significance. The degrees of freedom for the test statistic are the residual degrees of freedom in the ANOVA table.

Q3. [15, 18–19]
(a) The half-sib heritability is $4t_S$, as the sire variance component is a quarter of the additive genetic variance. From Question 2(c), $t_S = 0.103$ and the standard error of $t_S = 0.074$. The half-sib heritability and its standard error are 0.41 and 0.30, respectively.
(b) The maternal half-sib heritability is $4t_d$, which is a biased estimate of the heritability, as the dam variance component includes the maternal variance, a quarter of the dominance variance and the common environmental variance. From Question 2(c), $t_d = 0.207$, which has a standard error of 0.089, such that the maternal half-sib heritability is 0.83, with a standard error of 0.36.
(c) The full-sib heritability, $2(t_s + t_d)$, is also a biased estimate of the heritability. The standard error of the full-sib heritability is calculated from $2\sqrt{\text{var}(\sigma_s^2) + \text{var}(\sigma_d^2)}/\sigma_P^2$. The full-sib heritability and its standard error are 0.62 and 0.23.
(d) The maternal effect, c^2, is $t_s - t_d$ and the standard error is one half of the standard error of the full-sib heritability. The maternal effect and its standard error are 0.10 and 0.12.

Q4. [19]
The standard error of the heritability is estimated from the variance of the intra-sire correlation, equal to

$$\frac{2(1-t_{HS})^2(1+(n-1)t_{HS})^2}{(s-1)n(n-1)}$$

since the intra-sire correlation is $0.25h^2$. The standard error of the heritability estimated from data on half-sibs is 0.116.

Q5. [22–27]
(a) Let D denote the dam's litter size and d the average litter size of 30 daughters. The sums of squares and cross-products are

$$\Sigma D^2 = 1106, \quad \Sigma d^2 = 994.43, \quad \text{and} \quad \Sigma Dd = 1039.8$$

Estimates of variance and covariance components are

$$\sigma_D^2 = \frac{1}{9}\left[\Sigma D^2 - (\Sigma D)/10\right] = 2.711 \quad \sigma_d^2 = 0.047$$

and

$$\sigma_{Dd} = \text{cov}(D,d) = \frac{1}{9}\left[\Sigma Dd - (\Sigma D)(\Sigma d)/10\right] = 0.324$$

The regression coefficient $b_{dD} = \sigma_{dD}/\sigma_d^2 = 0.12$, and the intercept of $a = \bar{d} - b\bar{D} = \frac{1}{10}[\Sigma d - b\Sigma D] = 8.72$, indicate that the regression equation for the average litter size of 30 daughters on dam litter size is
8.72 + 0.12 dam litter size.

(b) The correlation between dam litter size and average litter size of 30 daughters is

$$r = \frac{\sigma_{Dd}}{\sigma_D \sigma_d} = 0.91$$

(c) The standard error of the regression coefficient is 0.019. The parameter z has a standard error of 0.378. Transformation of the confidence limits for z provides the 0.95 confidence interval of the correlation coefficient, of (0.65, 0.98).

(d) Calculation of the heritability from offspring–parent regression requires the covariance between offspring and parent. In the question, it was the average litter size of 30 daughters that was regressed on dam litter size. The covariance between the average litter size of 30 daughters and the dam's litter size is equal to the covariance between a daughter's litter size and the dam's litter size. The heritability is equal to twice the regression coefficient, which is 0.24.

The variance of the average litter size of 30 daughters is equal to 1/30 of the variance of a daughter's litter size, such that the variance of a daughter's litter size is 1.404. The correlation coefficient between daughter and dam litter sizes is $\frac{0.324}{\sqrt{1.404 \times 2.711}} = 0.166$. The heritability is twice the correlation coefficient, which is 0.33.

Q6. [28–29]
(a) The sire, dam and residual variance components for average daily gain are 13.75, 27.75 and 92.5, with the corresponding variance components for food conversion ratio equal to 55.0, 78.0 and 325.0. The phenotypic

variances for average daily gain and food conversion ratio are 134.0 and 458.0.

The sire, dam and residual covariance components are calculated in a similar manner to the variance components, with values equal to −19.5, −25.5 and −121.0, respectively. The phenotypic covariance is −166.0.

(b) The phenotypic correlation is $\sqrt{\dfrac{-166.0}{134.0 \times 458.0}} = -0.670$.

The additive genetic variances and covariance are four times the corresponding sire variances and covariance, such that the genetic correlation is

$$\sqrt{\dfrac{-78.0}{55.0 \times 220.0}} = -0.709.$$

The environmental correlation can be derived using the formula for the expectation of the phenotypic correlation:

$$r_P = r_A h_X h_Y + r_E \sqrt{(1 - h_X^2)(1 - h_Y^2)}$$

The environmental correlation is −0.641.

(c) The formula for the standard error of the genetic correlation is given in the text. The heritability of average daily gain and its standard error were determined in Question 3, and were equal to 0.412 and 0.312. For food conversion ratio the heritability and its standard error were 0.480 and 0.324. The standard error of the genetic correlation is 0.251.

Q7. [18]

In large litters there will be a larger number of animals competing for limited nutrients than in smaller litters, such that animals born into a large litter may be smaller than animals born into a small litter. Secondly, if smaller females have lower ovulation rate than larger females, then smaller females may have smaller litters than larger females. Therefore, females reared in larger litters may be smaller and subsequently have smaller litters than females reared in smaller litters.

The full-sib correlation in experiment 1, 0.054, is considerably lower than in experiment 2, 0.081, as standardisation of litter size will have reduced the similarity between full-sibs from larger litters, due to some litter mates being reared in different litters.

In experiment 2, the daughter–dam regression was substantially lower than the full-sib correlation, −0.028 versus 0.081, which suggests that the negative maternal environmental effect, of a daughter from a large litter having a small litter, was larger than the common environmental effect.

The full-sib correlation in experiment 1 was similar to the daughter–dam regression, 0.054 versus 0.049, which suggests that the common environmental effect was small.

Answers

Q8. [10, 31–32]

(a) The within-sow variance is the within-sow mean square of 3.23 and the between-sow variance is

$$\sigma_B^2 = (MS_B - MS_W)/n = (25.56 - 3.23)/10 = 2.23$$

The repeatability $r_e = \dfrac{\sigma_B^2}{\sigma_B^2 + \sigma_W^2} = 0.41$.

(b) The heritability for the mean of n measurements is

$$\dfrac{h^2}{r_e + (1 - r_e)/n}$$

For n = 2, 3 and 4, the proportional increase in the heritability is 0.42, 0.65 and 0.80.

Q9. [39–41]

(a and b)

Proportion selected	Selection differential	Response, $ih^2\sigma_P$	Response, $\dfrac{ih^2\sigma_P}{\sqrt{h^2 + (1-h^2)/n}}$
0.25	1.27	1.08	1.68
0.50	0.80	0.68	1.06
0.75	$0.424 = i_{(1-0.75)}\dfrac{(1-0.75)}{0.75}$	0.36	0.56

With five measurements per animal, the response was increased to 1.56 times the response with one measurement per animal.

Q10. [39–41]

The response $R = ib_{AS}\sigma_S = i\dfrac{\sigma_{AS}}{\sigma_S^2}\sigma_S = i\dfrac{\sigma_{AS}}{\sigma_S\sigma_A}\sigma_A = ir_{AS}\sigma_A$

The response $R = ir_{AS}\sigma_A = i\dfrac{b_{AS}}{b_{AS}}\dfrac{\sigma_{AS}}{\sigma_A\sigma_S}\sigma_A = i\dfrac{b_{AS}\sigma_{AS}}{b_{AS}\sigma_S}$

Since $b_{AS}\sigma_{AS}$ is the variance of the predicted breeding value, $\sigma_{\hat{A}}^2$, and $b_{AS}\sigma_S$ is the standard deviation of the predicted breeding value, then $R = i\sigma_{\hat{A}}$.

A second derivation sets $b_{AS}\sigma_{AS}$ equal to the covariance between the true and predicted breeding values, such that

$$R = i\dfrac{\sigma_{A\hat{A}}}{\sigma_{\hat{A}}} = i\dfrac{\sigma_{A\hat{A}}}{\sigma_{\hat{A}}^2}\sigma_{\hat{A}} = ib_{A\hat{A}}\sigma_{\hat{A}} = i\sigma_{\hat{A}}$$

because the regression of true breeding value on predicted breeding value is one, as the covariance between the true and predicted breeding values is the variance of the predicted breeding value.

Q11. [38–42]
(a) The predicted breeding values,

$$\hat{A} = \left[\frac{nh^2}{1+(n-1)r_e}\right](\bar{P} - \bar{P}_{Pop})$$

of animals A and B are 45.45 and 46.15. Although the average value for animal A is higher than that for animal B, the predicted breeding value for animal B is higher than that for animal A, as the smaller number of records for animal A result in a lower regression coefficient (0.454 versus 0.577).

(b) The variances of the predicted breeding values, $\text{var}(\hat{A}) = \dfrac{nh^4}{1+(n-1)r_e}\sigma_P^2$, of animals A and B are 31.82 and 40.38.

(c) The accuracies of the predicted breeding values,

$$r_{A\hat{A}} = \sqrt{\frac{nh^2}{1+(n-1)r_e}}$$

of animals A and B are 0.67 and 0.76. Note that $r_{A\hat{A}}^2 = b_{A\bar{P}}$.

(d) The prediction error variances, $\text{PEV} = (1 - r_{A\hat{A}}^2)\sigma_A^2$, of animals A and B are 38.18 and 29.62. The prediction error variance is another measure of the precision with which the breeding value is predicted, and so a small predicted error variance is desirable.

(e) The 95% confidence intervals of the predicted breeding values, $\hat{A} \pm 1.96\sqrt{\text{PEV}}$, for animals A and B are (33.44, 57.56) and (35.48, 56.82).

Q12. [44–48]
(a) Selection is on between-family deviations, as the individual is included in the sib mean, so the regression coefficient is calculated from

$$b_{A\bar{F}} = \left[\frac{1+(n-1)r}{1+(n-1)t}\right]h^2$$

where r =0.25 as the animals are half-sibs and t=0.10. The heritability is four times the half-sib correlation. The regression coefficient is 0.571 and the predicted breeding value of each of the five half-sibs is $0.571(\bar{F} - \bar{P}_{Pop})$.

(b) The same equation applies for the regression coefficient as in part (a), but with r = 0.5 as the animals are full-sibs and t = 0.36. The regression coefficient is 0.492.

(c)   Selection is on sib information, as the individual is excluded from the sib mean, so the regression coefficient is calculated from

$$b_{A\overline{S}} = \frac{nrh^2}{1+(n-1)t}$$

where $r = 0.5$, $t = 0.36$ and $n = 4$. The regression coefficient is 0.385 and the predicted breeding value of the individual is $0.385(\overline{S} - \overline{P}_{Pop})$.

(d)   Selection is on within-family deviations, so the regression coefficient is calculated from $\frac{(1-r)}{(1-t)}h^2$. The regression coefficient is 0.312 and the predicted breeding value of an individual is $0.312(P - \overline{F})$.

Q13. [49]
The response with family information relative to selection on the individual's measurement, ignoring differences in the selection intensity, is $\frac{b_{A\overline{F}}}{h^2} \frac{\sigma_{\overline{F}}}{\sigma_P}$.

		$\frac{b_{A\overline{F}}}{h^2}$	$\frac{\sigma_{\overline{F}}}{\sigma_P}$	Relative response
(a)	Between-family half-sib	$\frac{1+(n-1)r}{1+(n-1)t}$	$\sqrt{t+\frac{1-t}{n}}$	
		1.428	0.529	0.756
(b)	Between-family full-sib	1.230	0.698	0.859
(c)	Sib information	$\frac{nr}{1+(n-1)t}$	$\sqrt{t+\frac{1-t}{n}}$	
		0.962	0.721	0.693
(d)	Within-family	$\frac{1-r}{1-t}$	$\sqrt{\left(1-\frac{1}{n}\right)(1-t)}$	
		0.781	0.716	0.559

In each case, individual selection had a relatively larger response, ignoring differences in the selection intensities.

Q14. [44–48]

	Selection method	Predicted breeding value A	B	Select	Calculation of breeding value for animal A
(b)	Between-family with full-sibs	−14.75	19.67	B	0.492 (920–950)
(c)	Sib information	−17.31	17.31	B	0.385 (905–950) Full-sib mean excluding the individual = 905
(d)	Within-family	18.75	−6.25	A	0.312 (980–920)
	Individual only	12.00	8.00	A	h^2 (980–950)

Animal B was selected, with selection on between-family deviations or selection on sib information as the full-sib mean of animal B was larger than the full-sib mean of animal A. In contrast, animal A had a larger within-family deviation than animal B.

Q15. [50]

The predicted breeding values, $\dfrac{2nh^2}{4+(n-1)h^2}(\bar{P}-\bar{P}_{Pop})$, of bulls A and B are 461.5 and 400.0. The progeny mean of bull B is regressed back to the population mean to a greater extent than the progeny mean of bull A, due to the lower number of progeny.

Q16. [68–71]
(a) The heritability of average daily gain is four times the half-sib correlation, which equals 0.4. The value of r is 0.25, the genetic relationship between the individual and its half-sibs. The a matrix is equal to one and the C matrix is the genetic variance, as the trait in the selection criterion is the same as in the selection objective. The P and G matrices are $\begin{bmatrix} 1.00 & 0.10 \\ 0.10 & 0.28 \end{bmatrix}\sigma_P^2$ and $\begin{bmatrix} 0.40 \\ 0.10 \end{bmatrix}\sigma_P^2$, with selection criterion coefficients of $b = \begin{bmatrix} 0.378 \\ 0.222 \end{bmatrix}$.

An animal's predicted breeding value is

$$\hat{A} = 0.378(P-\bar{P}_{Pop}) + 0.222(\bar{S}-\bar{P}_{Pop})$$

The accuracy of predicted genetic merit, $r_{IH} = \sqrt{\dfrac{b'\,Pb}{a'\,Ca}}$, is 0.658.

(b) The phenotypic variance of the individual's deviation from the mean of its sibs, when the individual is excluded from the sib mean, is

$$\text{var}(\text{ind}-\overline{S}) = \text{var}(\text{ind}) - 2\text{cov}(\text{ind},\overline{S}) + \text{var}(\overline{S}) = \left[1 - 2t + t + \frac{1-t}{n}\right]\sigma_P^2$$

The phenotypic covariance between the individual's deviation from the mean of its sibs and the sib mean is

$$\text{cov}(\text{ind}-\overline{S},\overline{S}) = \text{cov}(\text{ind},\overline{S}) - \text{var}(\overline{S}) = \left[t - \left(t + \frac{1-t}{n}\right)\right]\sigma_P^2$$

The genetic covariance between the individual and the individual's deviation from the mean of its sibs is $h^2 - rh^2$, and the genetic covariance between the individual and the mean of its sibs is rh^2.

The P and G matrices are $\begin{bmatrix} 1.08 & -0.18 \\ -0.18 & 0.28 \end{bmatrix}\sigma_P^2$ and $\begin{bmatrix} 0.30 \\ 0.10 \end{bmatrix}\sigma_P^2$, with the selection criterion coefficients equal to $b = \begin{bmatrix} 0.378 \\ 0.600 \end{bmatrix}$.

The accuracy of a predicted breeding value is 0.658.

(c) The phenotypic covariance of the individual's measurement and the family mean is the variance of the family mean, as

$$\text{cov}(\text{ind},\overline{F}) = \frac{1}{n}[\text{var}(\text{ind}) + (n-1)\text{cov}(\text{ind},\text{sib})] = \left[\frac{1}{n} + \frac{n-1}{t}\right]\sigma_P^2$$

Similarly, the genetic covariance of the individual's measurement and the family mean is $\left[\frac{1}{n} + \frac{n-1}{n}r\right]h^2\sigma_P^2$. The P and G matrices are $\begin{bmatrix} 1.00 & 0.25 \\ 0.25 & 0.25 \end{bmatrix}\sigma_P^2$ and $\begin{bmatrix} 0.40 \\ 0.15 \end{bmatrix}\sigma_P^2$, with selection criterion coefficients of $b = \begin{bmatrix} 0.333 \\ 0.267 \end{bmatrix}$.

The accuracy of a predicted breeding value is 0.658.

A point worth noting regarding parts (a), (b) and (c) of the question is that the selection criteria use the same information, which is the reason for the same accuracies of the predicted breeding values.

In part (a) the selection criterion is $b_1\text{ind} + b_2\overline{S}$ and in part (b) the selection criterion is

$$b_3(\text{ind}-\overline{S}) + b_4\overline{S} = b_3\text{ind} + (b_4-b_3)\overline{S}$$

In particular, the selection criterion coefficient for the sib mean in part (a) is the difference between the two selection criterion coefficients in part (b). In part (c), the selection criterion is $b_5\text{ind} + b_6(\text{ind}+5\overline{S})/6$ which is equal to $(b_5 + b_6/6)\text{ind} + (5b_6/6)\overline{S}$, such that b_5 and b_6 are functions of b_1 and b_2, with $b_1 + b_2 = b_5 + b_6$.

(d) The selection criterion coefficients are 0.344 and 0.156 for part (a), 0.344 and 0.500 for part (b) and 0.312 and 0.188 for part (c).

Q17. [72]

The phenotypic and genetic variance–covariance matrices, P and G, are $\begin{bmatrix} 1.232 & 2.285 \\ 2.285 & 6.150 \end{bmatrix}$ and $\begin{bmatrix} 0.641 & 0.891 \\ 0.891 & 2.460 \end{bmatrix}$, with selection criterion coefficients of $b = \begin{bmatrix} 0.514 \\ 1.954 \end{bmatrix}$, given the vector of economic weights, $a = \begin{bmatrix} 1 \\ 5 \end{bmatrix}$. The selection criterion is 0.514 ADG + 1.954 FCR. For selection purposes, the mean values for ADG and FCR need not be subtracted from the measurements, but the predicted breeding values are determined from $0.514(\text{ADG} - \overline{\text{ADG}}) + 1.954(\text{FCR} - \overline{\text{FCR}})$.

Q18. [63–64]

The equation for the correlated responses is $i \dfrac{b' G_j}{\sqrt{b' Pb}}$, where i is the standardised selection differential and G_j is the jth column of matrix G, corresponding to trait j. For average daily gain, the vector G_j is $\begin{bmatrix} 0.641 \\ 0.891 \end{bmatrix}$. The correlated responses in average daily gain and food conversion ratio are 0.388i and 0.988i. The equation for the correlation between a trait in the selection objective with the selection criterion is $\dfrac{b' G_j}{\sqrt{b' Pb.C_{jj}}}$, where C_{jj} is the variance of trait j. The correlations for average daily gain and food conversion ratio with the selection criterion are 0.486 and 0.630.

Q19. [61]

The phenotypic variance of the selection criterion is b'Pb and the genetic variance of the selection criterion is b'Gb. The heritability of the selection criterion is 0.40.

Q20. [73]

The P matrix is the same as in Question 17, but the G matrix is replaced by the column of G in Question 17 corresponding to average daily gain, $\begin{bmatrix} 0.641 \\ 0.891 \end{bmatrix}$. The selection criterion coefficients are $b = \begin{bmatrix} 0.808 \\ -0.155 \end{bmatrix}$ and the correlated responses in average daily gain and food conversion ratio are 0.616i and 0.550i, respectively.

The selection criterion coefficients and correlated responses to selection could have been obtained by using the P and G matrices in Question 17, but changing the economic value of food conversion ratio to zero, such that $a = \begin{bmatrix} 1 \\ 0 \end{bmatrix}$.

The response to selection on average daily gain would be $ih^2\sigma_P = 0.577i$, so inclusion of food conversion ratio in the selection criterion only improved the response in average daily gain by 7%.

Q21. [86–89]
When there are several traits, setting up the phenotypic and genetic (co)variance matrices can be very time consuming if each element is calculated separately. Matrix multiplication can be used to generate the phenotypic and genetic (co)variance matrices. For example, the phenotypic covariance between traits X and Y is $\text{cov}_P(X,Y) = r_P \sigma_X \sigma_Y = \sigma_X r_P \sigma_Y$ and, similarly, the genetic covariance can be written as $\text{cov}_A(X,Y) = r_A h_X h_Y \sigma_X \sigma_Y = \sigma_X h_X r_A h_Y \sigma_Y$. Therefore, if S is a diagonal matrix, consisting of the phenotypic standard deviations of each trait, H is a diagonal matrix consisting of the square root of the heritability for each trait, and Rp and R_A are the phenotypic and genetic correlation matrices, then the phenotypic and genetic (co)variance matrices can be written as SR_PS and SHR_AHS, respectively.

The matrices are
$$S = \begin{bmatrix} 0.102 & 0 & 0 \\ 0 & 3.48 & 0 \\ 0 & 0 & 4.0 \end{bmatrix}, \quad R_P = \begin{bmatrix} 1.0 & 0.06 & 0 \\ 0.06 & 1.0 & 0 \\ 0 & 0 & 1.0 \end{bmatrix}$$

$$R_A = \begin{bmatrix} 1.0 & -0.05 & 0 \\ -0.05 & 1.0 & -0.6 \\ 0 & -0.6 & 1.0 \end{bmatrix} \text{ and } H = \begin{bmatrix} \sqrt{0.46} & 0 & 0 \\ 0 & \sqrt{0.31} & 0 \\ 0 & 0 & \sqrt{0.45} \end{bmatrix}, \text{ and the}$$

genetic (co)variance matrix is $\begin{bmatrix} 0.048 & -0.0067 & 0.0 \\ -0.0067 & 3.754 & -3.119 \\ 0.0 & -3.119 & 7.200 \end{bmatrix}$. The P, G and C matrices are just submatrices of the phenotypic and genetic (co)variances. For example, the P and G matrices with one backfat measurement are

$$P = \begin{bmatrix} & \text{ADG} & \text{BFAT} \\ \text{ADG} & 0.0104 & 0.0213 \\ \text{BFAT} & 0.0213 & 12.110 \end{bmatrix} \text{ and } G = \begin{bmatrix} & \text{ADG} & \text{LEAN} \\ \text{ADG} & 0.048 & 0.0 \\ \text{BFAT} & -0.0067 & -3.119 \end{bmatrix}$$

The element of P corresponding to the variance of the repeated backfat depth measurements is $\left[r_e + \dfrac{1-r_e}{n}\right]\sigma^2_{\text{BFAT}}$. The correlations between the selection objectives and the selection criteria are 0.426, 0.449, 0.463 and 0.468, when the mean of one, two, four and six backfat measurements are included in the selection criterion. The increase in the correlation is due to the reduction in the variance of the mean backfat measurement.

Q22. [72]

The phenotypic and genetic (co)variance matrices can be calculated using the procedure outlined in the answer to Question 21, such that with the selection criterion consisting of ADG, BFAT and ADG

$$P = \begin{bmatrix} 0.0104 & 0.0213 & 0.0122 \\ 0.0213 & 12.110 & 0.323 \\ 0.0122 & 0.323 & 0.0488 \end{bmatrix} \text{ and } G = \begin{bmatrix} 0.0048 & 0.0 \\ -0.0067 & -3.119 \\ 0.0052 & -0.156 \end{bmatrix}$$

for the selection objective consisting of ADG and LEAN.

The selection criterion is 680 ADG −12.6 BFAT −174 ADF.

The correlated responses in each trait are obtained from $i \dfrac{b' G_j}{\sqrt{b' P b}}$, where i is the standardised selection differential and G_j is the jth column of the genetic (co)variance matrix, which corresponds to trait j, after omitting the row for carcass lean content, which is not included in the selection criterion. For example, transpose of the vector G_j for ADF is [0.0052 0.1348 0.0166]. The correlated responses are 31 g/day, −0.94 mm, −13 g/day and 8.3 g/kg for average daily gain, backfat depth, daily food intake and carcass lean content, per standardised selection differential.

Question 21 refers to when average daily food intake was not included in the selection criterion.

The selection criterion is 485 ADG −16.8 BFAT and the correlated responses, per standardised selection differential, are 33 g/day, −0.89 mm, 3 g/day and 7.1 g/kg for average daily gain, backfat depth, daily food intake and carcass lean content. The economic value of the response, per standardised selection differential, of 74.4, is lower than 79.8, when average daily food intake was included in the selection criterion.

Q23. [75–78]

The new P matrix, equal to $\begin{bmatrix} P & G_{ADF} \\ G'_{ADF} & 0 \end{bmatrix}$, is obtained by adding the row and column of the G matrix, which corresponds to daily food intake and setting the remaining diagonal term of NP to zero. The new G matrix, $\begin{bmatrix} G \\ \underline{0} \end{bmatrix}$, is obtained by adding a row of zeros, $\underline{0}$.

The selection criterion is 710 ADG −11.93 BFAT −124 ADF.

The correlated responses, per standardised selection differential, are 36 g/day, −0.84 mm, 0 g/day and 7.2 g/kg for average daily gain, backfat depth, daily food intake and carcass lean content.

The correlation between the selection objective and criterion, when the response in daily food intake is constrained to zero, is 0.450, which is marginally lower than the correlation of 0.457, when no restriction was imposed on the response in daily food intake. The economic value of the response, per standardised selection differential, is 78.6, which is similar to when no restriction was placed on the response in daily food intake.

Q24. [87–89]
For one trait, the phenotypic covariance between the individual's measurement and the sib mean is equal to the variance of the sib mean, as

$$\text{cov}(\text{ind}, \overline{\text{sib}}) = \frac{1}{n}[\text{var}(\text{ind}) + (n-1)\text{cov}(\text{ind}, \text{sib})] = \frac{1}{n}[1 + (n-1)t]\sigma_P^2$$

The phenotypic covariance between the individual's measurement for ADG and BFAT is

$$\text{cov}(\text{ind}_{ADG}, \overline{\text{sib}}_{BFAT}) = \frac{1}{n}\left[r_P + (n-1)\frac{1}{2}r_A h_{ADG} h_{BFAT}\right]\sigma_{ADG}\sigma_{BFAT}$$

which has the same structure as the phenotypic covariance between the individual's measurement and the sib mean for the same trait, but with t replaced by $\frac{1}{2}r_A h_{ADG} h_{BFAT}$, assuming that the common environmental covariance between traits is zero. The genetic covariances between sib mean for average daily gain with the individual's genetic merit for ADG and LEAN are

$$\begin{bmatrix} \frac{1}{n}\left(1 + (n-1)\frac{1}{2}h_{ADG}^2\right)\sigma_{ADG}^2 \\ \frac{1}{n}\left(1 + (n-1)\frac{1}{2}r_A h_{ADG} h_{LEAN}\right)\sigma_{ADG}\sigma_{LEAN} \end{bmatrix}$$

with corresponding values for the sib mean for backfat depth.
With measurements on the individual and three full-sibs, the P and G matrices are

$$P = \begin{bmatrix} 0.0104 & 0.0213 & 0.0049 & 0.0028 \\ 0.0213 & 12.110 & 0.0028 & 4.9350 \\ 0.0049 & 0.0028 & 0.0049 & 0.0028 \\ 0.0028 & 4.9350 & 0.0028 & 4.9350 \end{bmatrix} \text{ and } G = \begin{bmatrix} 0.0048 & 0.0 \\ -0.0067 & -3.1194 \\ 0.0030 & 0.0 \\ -0.0042 & -1.9497 \end{bmatrix}$$

Note that the P matrix consists of three identical 2 × 2 submatrices, since the covariance between the individual's measurement and the full-sib mean is the variance of the full-sib mean, when the individual is included in the full-sib mean.

The selection criterion coefficients and the accuracy of selection for different numbers of full-sibs are as follows:

		Individual	Number of full-sibs (excluding the individual)			
			1	2	3	4
Accuracy	r_{IH}	0.43	0.44	0.45	0.45	0.46
Selection criterion coefficients	growth	485	354	354	354	354
	$\overline{\text{growth}}$		197	236	263	281
	backfat	−16.8	−11.0	−11.0	−11.0	−11.0
	$\overline{\text{backfat}}$		−9.5	−12.0	−13.8	−15.3

There is little gain, in this instance, from including full-sib information in the selection criterion. If there is no extra cost in obtaining the full-sib measurements, then the additional measurements should be used to predict the individual's breeding value. Note that when the individual's measurements are included in the full-sib mean, then the selection criterion coefficients corresponding to the individual's measurements are constant, irrespective of the number of full-sibs.

Q25. [94–95]

When $r_A h_X = r_P h_Y$, then the response, R, in trait Y with selection on traits X and Y is proportional to h_Y^2, but the actual response, R^*, is proportional to $\dfrac{h_Y^4}{\sqrt{h_Y^4 + \Delta}}$, where $\Delta = \dfrac{\alpha^2 h_X^2 h_Y^2}{1 - r_P^2}$, given that the selection criterion was determined using the estimate of the genetic correlation, $r_A + \alpha$. The predicted response, $\hat{R}$, is proportional to $\sqrt{h_Y^4 + \Delta}$.

α	Δ	Actual loss, $\dfrac{R - R^*}{R}$ (%)	Predicted gain, $\dfrac{\hat{R} - R}{R}$ (%)
−0.1	0.000625	2.98	3.08
0.2	0.0025	10.56	11.80

The actual loss in response is essentially equal to the predicted gain from including information on trait X in the selection criterion. Note that the above equations are valid only when $r_A h_X = r_P h_Y$.

Q26. [98–99]

Annual rate of response (R) = $\dfrac{\left(i_M r_{IH_M} + i_F r_{IH_F}\right)}{L_M + L_F} \sigma_A$. In Programme A $i_M = 2.421$ and $i_F = 0.195$, such that $R = 0.141 \sigma_A$, while in Programme B, $i_M = 1.755$ and $R = 0.155 \sigma_A$. The longer generation interval of Programme B is compensated for by the higher accuracy of selection, relative to Programme A.

Q27. [118–120]

The three animals, A, B and C, are unrelated, so the P matrix consists of nine 3 × 3 diagonal submatrices. The phenotypic variances for measurements on the animal, its progeny and its sire are $\left[r_e + \dfrac{1 - r_e}{n}\right]\sigma_P^2$, $\left[t + \dfrac{1 - t}{n}\right]\sigma_P^2$ and σ_P^2, respectively. The phenotypic covariance between the animal and the mean measurement of its progeny and with its sire's measurement is $0.5h^2\sigma_P^2$, while the phenotypic covariance between the animal's sire and the mean measurement of the animal's progeny is $0.25h^2\sigma_P^2$. Similarly, the G matrix consists of three diagonal 3x3 submatrices, with elements σ_A^2, $0.5\sigma_A^2$ and $0.5\sigma_A^2$, respectively. The "selection criterion" for the predicted breeding value of animal B is

$$0.286 \; \overline{\text{animal}} + 0.952 \; \overline{\text{progeny}} + 0.048 \; \overline{\text{sire}}$$

The predicted breeding values for animals A, B and C are 45.7, 53.3 and 41.4, with prediction error variances of 7.1, 8.6 and 10.9 and accuracies of 0.91, 0.89 and 0.85, respectively.

Predicted breeding values can be determined separately for each animal, since animals A, B and C are unrelated. The P and G matrices for animal B are $\begin{bmatrix} 70 & 20 & 20 \\ 20 & 14.5 & 10 \\ 20 & 10 & 100 \end{bmatrix}$ and $\begin{bmatrix} 40 \\ 20 \\ 20 \end{bmatrix}$.

Q28. [118–120]
Calculation of the relationships between animals is required to determine the P matrix, for animals with measurements, and the G matrix, for animals with and without measurements:

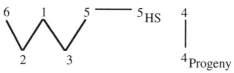

Each of the lines in the diagram represents $\frac{1}{2}h^2\sigma_P^2$, except for the line between animal 5 and its half-sibs, which is equal to $\frac{1}{4}h^2\sigma_P^2$. The P matrix represents animal 1, the progeny of animal 4, animal 5's half-sibs and animal 6. The off-diagonal elements of P are equal to zero and the diagonal elements of P are $\left[1 \quad t+\dfrac{1-t}{10} \quad t+\dfrac{1-t}{10} \quad 1 \right]\sigma_P^2$. The G matrix is the covariance between animals with measurement and all animals in the pedigree, with

$$G = \begin{array}{r} \\ \\ 4'\text{s progeny} \\ 5'\text{s half-sibs} \\ 6 \end{array} \begin{bmatrix} 1 & 4 & 5 & 6 & 2 & 3 \\ 1 & 1 & 0 & 0 & 0 & 0.5 & 0.5 \\ 0 & 0.5 & 0 & 0 & 0 & 0 \\ 0 & 0 & 0.25 & 0 & 0 & 0.125 \\ 0 & 0 & 0 & 1 & 0.5 & 0 \end{bmatrix} \sigma_A^2$$

The predicted breeding values, their accuracies and prediction error variances for the six animals are as follows:

Animal	1	2	3	4	5	6
Predicted breeding value	8	10	9	20	10	12
Prediction error variance	48	64	69.37	37.89	69.47	48
Accuracy	0.63	0.45	0.36	0.72	0.36	0.63

The prediction error variances and accuracies reflect the amount of information available on each animal. Animals 1 and 6 each have one record. Information from the half-sibs of animal 5 is used for animals 3 and 5, but there is also information from animal 1 for animal 3. Animal 4 has the most information, with ten progeny.

Q29. [125–127]
Information on the number of observations in each herd and for each sire is given in the X'X and the Z'Z diagonal matrices, the diagonals of which are $[3\ 3\ 3]$ and $[2\ 3\ 4]$. The X'Z matrix describes the number of observation per herd–sire class,

$$\begin{bmatrix} 1 & 1 & 1 \\ 1 & 2 & 0 \\ 0 & 0 & 3 \end{bmatrix}$$

The sum of observations is 490, 420 and 300 for herds and 390, 400 and 420, respectively. The sires are unrelated, so the numerator relationship matrix is equal to the identity matrix. Given a heritability of 0.4 and measurements on progeny, then $\lambda = \dfrac{4-h^2}{h^2} = 9$. The mixed model equations are

$$\begin{bmatrix} 3 & 0 & 0 & 1 & 1 & 0 \\ 0 & 3 & 0 & 1 & 2 & 0 \\ 0 & 0 & 3 & 1 & 0 & 3 \\ 1 & 1 & 1 & 9 & 0 & 0 \\ 1 & 2 & 0 & 0 & 9 & 0 \\ 0 & 0 & 3 & 0 & 0 & 9 \end{bmatrix} \begin{bmatrix} \hat{b} \\ \hat{u} \end{bmatrix} = \begin{bmatrix} 490 \\ 420 \\ 300 \\ 390 \\ 400 \\ 420 \end{bmatrix}$$

When the relationships between sires are accounted for, the numerator relationship matrix, A, is

$$\begin{bmatrix} 1 & 0.25 & 0.125 \\ 0.25 & 1 & 0.5 \\ 0.125 & 0.5 & 1 \end{bmatrix}$$

Q30. [130–133]

(a) For a heritability of 0.3 and a repeatability of 0.5, $\lambda = \dfrac{1-r_e}{h^2} = 1.67$ and

$\gamma = \dfrac{1-r_e}{r_e - h^2} = 2.5$. The mixed model equations are

$$\begin{bmatrix} 5 & 0 & 3 & 2 & 3 & 2 \\ 0 & 4 & 1 & 3 & 1 & 3 \\ 3 & 1 & 6.733 & -0.267 & 4 & 0 \\ 2 & 3 & -0.267 & 7.733 & 0 & 5 \\ 3 & 1 & 4 & 0 & 6.5 & 0 \\ 2 & 3 & 0 & 5 & 0 & 7.5 \end{bmatrix} \begin{bmatrix} \hat{b} \\ \hat{u}_A \\ \hat{u}_r \end{bmatrix} = \begin{bmatrix} 101 \\ 80 \\ 77 \\ 104 \\ 77 \\ 104 \end{bmatrix}$$

(b) For each animal, two effects are estimated, u_A and u_r, which provide information on the animal's genetic merit, u_A, and the animal's future performance $u_A + u_r$.

(c) Prediction error variances of predicted breeding values are the diagonal elements of $C^{22}(1-r_e)\sigma_P^2$ submatrix, determined from the inverse of the left-hand side matrix (LHS) in the mixed model equations, with LHS^{-1} equal to

$$\begin{bmatrix} C^{11} & C^{12} & C^{13} \\ C^{21} & C^{22} & C^{23} \\ C^{31} & C^{32} & C^{33} \end{bmatrix}$$

Prediction error variances are used to calculate the accuracy and variance of a predicted breeding value and the effective number of records on animals.

and the 3×3 submatrix with diagonal elements equal to 9, is replaced by λA^{-1}.

Index

Accuracy 40
    measurements on progeny 54
    measurements on sibs 54
    predicted breeding value 110, 111, 133
    predicted genetic merit 60
    regression 40
    repeated measurements 54
    response 41, 54, 55
Additive genetic effect 17, 130
Additive genetic variance 133, 145
Analysis of covariance 28
Analysis of variance 25
Analysis of variance (*see* ANOVA) 6
ANOVA
    balanced and unbalanced designs 7, 13
    between-dam variance 13
    between-group variance 6, 7, 8
    between-sire variance 13
    correction factor 6
    expectation of mean squares 7
    within-group variance 6, 7, 8

Bias
    heritability 19
    variances 16
Binary trait 155
    heritability 159
    incidence 156
    mixed model equations 162
    selection differential 156
BLUP 122

Common environmental effect 17, 137, 138
Confidence interval 42
Correlation 24, 40
    coefficient 24
    environmental 29
    full-sib 14
    genetic 29
    half-sib 14
    phenotypic 29
    standard error 24
Covariance 22
    additive genetic 28
    environmental 28
    maternal 28
    phenotypic 28
Covariance matrix
    genetic 59, 76

Degrees of freedom 7
Disequilibrium variance 100
Dominance deviation 144
Dominance effect 17

Economic value 90
Effective population size 102
Eigenvalues 96
Eigenvectors 96
Errors in parameter estimates
    contribution of traits 94
    response 93

Fixed effects 122, 126

Gene of large effect 146
General environmental effect 17
Generation interval 98, 106, 110
Genes of known large effect 143
    response 148
Genetic correlation 109, 140
    standard error 29
Genetic markers 150
Genetic relationship 45
Genotype with environment
    interaction 140

Heritability 19, 36, 39
    additive genetic variance 19
    bias 19
    binary trait 159
    half-sib correlation 19
    liability 156, 159
    offspring–parent correlation 27
    offspring–parent regression 27
    repeatability 31
    standard error 19
Heterozygote advantage 144

Inbreeding 101
    depression 101
    rate of 102
    reduction in 103
Incidence
    binary trait 156
Individual animal model 130, 134
Infinitesimal model 143

Liability 156
    heritability 156, 159
Logit function 161

Marker assisted selection 150
Maternal effect 19
Maternal genetic effect 17, 138
Matrix
    addition 173
    definition 171
    inverse 172
    multiplication 172

    subtraction 173
    transpose 173
    variance–covariance 173
Mean 2
    standard error 4
Mean squares 7
    expectation 7
Mendelian sampling 53, 146
Mixed model equations 123, 124, 129, 130, 132, 135, 141
    binary trait 162
MOET 110

Normal distribution 4, 37
Numerator relationship matrix 124, 127, 130, 131

Offspring–parent regression 27

Parameter estimates
    errors 93
    modification 95
Performance testing 106, 108
Phenotype 17
Phenotypic correlation
    standard error 29
Phenotypic variance 14
Predicted breeding value
    accuracy 40, 52, 109, 110, 133
    confidence interval 42
    parental information 52
    prediction error variance 41
    progeny measurements 50
    repeated measurements 35, 50
    sib measurements 44
    single measurement 34
    variance 39, 41, 133
Predicted breeding values
    variance–covariance matrix 115
Predicted genetic merit 62
    accuracy 60
Prediction error variance 41, 42, 133
Progeny testing 106, 108

Quantitative trait locus 150

Random effects 122, 126

Recombination 154
Regression
    offspring–parent 27
    standard error 25
REML 128, 140
Repeatability 10, 14, 30, 38, 129
    between-sire variance 31
    half-sib correlation 31
    heritability 31
    standard error 10
Repeated measurements 129
Response
    accuracy 54, 55
    between-family selection 49
    correlated traits 62
    errors in parameter estimates 93
    genes of known large effect 148
    long-term 99
    measurements on progeny 54, 107
    measurements on sibs 54
    multiple measurements 38
    progeny selection 51
    repeated measurements 54
    selection objective 62
    short-term 99
    single measurement 38
    within-family selection 49

Selection
    between-family deviations 46
    sib information 44
    within-family deviations 47
Selection criterion 59, 146
    animal 68
    animal and sibs 69
    coefficients 61
    contribution of traits 65, 73, 94
    variance 60
Selection differential 37, 110
    binary trait 156
    standardised 37, 49, 63, 98
Selection objective 59
    contribution of traits 65
    desired gains 79
    restricted 75
    two traits 72
    variance 60

Selection within-family 112
Sire model 123, 129, 134
Specific animal effect 130
Specific environmental effect 30
Standard deviation 2
Standard error
    correlation coefficient 24
    dam mean 14
    genetic correlation 29
    group mean 31
    heritability 19
    mean 4
    phenotypic correlation 29
    regression coefficient 25
    repeatability 10
    sire mean 14
    variance components 15

Threshold model 155

Variance 2
    additive genetic 18
    between-dam 13, 18
    between-group 6, 7, 8
    between-sire 13, 18
    bias 16
    disequilibrium 100
    environmental 18
    group mean 7
    maternal 18
    phenotypic 14, 18
    predicted breeding value 39, 41
    prediction error 41
    properties 23
    reduction in 99, 108
    residual 18
    selection criterion 60
    selection objective 60
    within-group 6, 7, 8
Variance–covariance matrix 173
    genetic 59, 114
    phenotypic 59, 76, 114
    predicted breeding values 115
    prediction error 115

Within-family selection 112